Volume II
Powder Coating

A Practical Guide to Equipment, Processes and Productivity at a Profit

Volume II
POWDER COATING

A Practical Guide to Equipment, Processes and Productivity at a Profit

Mike Cowley
Ad-Qual Group.

JOHN WILEY & SONS
CHICHESTER • NEW YORK • WEINHEIM • BRISBANE • TORONTO • SINGAPORE

Published in association with

SITA TECHNOLOGY LIMITED
LONDON, UK

SITA TECHNOLOGY

Thornton House
Thornton Road
London
SW19 4NG

Published in 1999 by
John Wiley & Sons Ltd
In association with SITA Technology Limited

Wiley Editorial Offices
John Wiley & Sons Ltd., Baffins Lane,
Chichester, West Sussex, England. PO19 1UD

John Wiley & Sons Inc., 605 Third Avenue,
New York, NY 10158-0012, USA

VCH Verlagsgellesellschaft mvH, Pappalallee 3
D-69469 Weinheim, Germany

Jacaranda Wiley Ltd, G.P.O Box 859, Brisbane
Queensland 4001, Australia

John Wiley & Sons (Canada) Ltd, 22 Worcester Road,
Rexdale, Ontario M9W 1L1, Canada

John Wiley & Sons (SEA) Pte Ltd, 37 Jalan Pemimpin Road 05-04,
Block B, Union Industrial Building, Singapore 2057

A catalogue record for this book is available from the British Library

ISBN 471 979007 1999

Printed and bound in the UK by Short Run Press Limited
Bittern Road, Sowton Industrial Estate, Exeter.

CONTENTS

Chapter I: INTRODUCTION **3-10**

History 4
Why use powder coatings? 6

The powder coating application process 7

Chapter II: ASSESSMENT OF THE COMPONENT BEING COATED **13-31**

Designing the component for coating 13
Sharp edges 13
Welds 14
Crimping 15
Rolled edges 16
Holes for bolts and screws 16
Threads 16
Components with moving parts 17
Allowing for coating thickness 17
The substrate 18
Metals 18
Ferrous metals 19
Non-ferrous metals 19
Galvanised components 19
Zinc-plated ferrous components 20
Glass 20
Plastics 20
Wood 21
Adhesion 21
Wetting contact theory 23
Mechanical adhesion 23
Chemical adhesion 24
Handling of components 24
During pre-treatment 25
After pre-treatment 25
During application 26
Conveyor loading 27

Cleaning of hooks and jigs 28
Incineration or pyrolysis 28
Fluidised bed cleaning 29
Chemical stripping 29
Mechanical stripping 30
Masking 30

Chapter III: PRE-TREATMENT 35-65

Why pre-treat? 36
Removing contamination 36
Methods of pre-treatment 38
Wipe cleaning 38
Mechanical pre-treatment 40
Brushing 40
Abrasive pads 40
Shot or grit blasting 41
Chemical cleaning 44
Vapour degreasing 45
Conveyorised tunnel systems 47
Ultrasonic cleaning 47
Spray cleaning 48
Power-wash cleaning 48
Steam cleaning 49
Alternative cleaning materials 50
Corrosion 52
Conversion coatings 53
Phosphating 56
Passivation of phosphate layers 58
Combined cleaning and conversion 59
Chromate conversion coatings 62
Effluent treatment 63
Other pre-treatments 64
Phosphoric acid phosphating 64
Pigmented etch primers 64
Primer coating powders 65
Electrophoretic primers 65

CHAPTER IV: POWDER COATING APPLICATION 69-115

The choice 69
Why coat? 70
What substrate? 70

How many items? 71
What is the specification? 71
How thick? 72
What colours? 72
Production factors? 72
Processing costs? 72
What will the equipment cost? 73
The final choice 73
Fluidised bed coating 75
Fluidised bed equipment 75
Water quenching 81
Post-heating and curing 81
Conveyorised fluidised bed coating 82
Electrostatic spraying 82
The theory of electrostatic powder coating 83
Corona charging 86
Tribocharging 89
Manual electrostatic application 91
Powder feed systems 93
Fluidised bed hopper 93
Box units 95
Injectors and venturis 97
Electrostatic powder spray guns 99
Conveyorised electrostatic fluidised bed coating 105
Flock spraying - with electrostatic spray guns 106
Automatic electrostatic spraying equipment 107
Air-assisted spray guns – corona charging 108
Automatic air-assisted atomisers – tribocharging 110
Powder bells 111
Powder discs 111
Automatic applicators 114

Chapter V: POWDER SPRAY BOOTHS 119-144

Health and safety 119
Materials of construction 120
Design 122
Cleaning 123
Selection of spray booth type 124
Manual booths 125
Automatic booths 128
Colour change 130
Change in polymer type 134

The importance of transfer efficiency 135
Powder recovery 136
Cyclone recovery 137
Filters 139
Cartridge filter systems 140
Powder recycling 142

Chapter VI: HEATING, MELTING AND CURING SYSTEMS **147-155**

Temperature, time and flow out 147
Box ovens 148
Conveyorised ovens 149
Convection ovens **151**
Radiation heating 152
Infrared ovens for pre-heating 154
Other types of oven 154
Induction heating 154
Ultraviolet curing 154
Electron beam curing 155

Chapter VII: POWDER COATING versus LIQUID PAINT **159-162**

Cost 159
Raw materials 159
Processing costs 160
Stoving, melting and curing costs 160
Equipment and other costs 161
Quality 161
Health, safety and the environment 162

Chapter VIII: POWDER COATING IN PRACTICE **165-168**

Metals 165
Glass 166
Wood 167
MDF 167
Plastics 167
Other materials 168
Rubber 168
Textiles 168
Food products 168

Chapter IX: MANAGING QUALITY **171-180**

Control of quality in powder coating 171
The FMECA technique 172
Auditing quality 176
Statistical process control 177
Adopting a quality assurance culture 179

Chapter X: TESTING OF POWDER COATINGS **183-191**

Introduction 183
Evaluation of coating powder 184
Tests during and after coating 184
Coating thickness assessment of electron statically applied powders before heating and curing 185
Colour 186
Gloss 186
Dry film thickness 187
Test panels 187
Adhesion 187
Cure 187
Flexibility 188
Troubleshooting and other quality issues 188
Air quality 188
Troubleshooting 188

Appendix I: HEALTH AND SAFETY AND ENVIRONMENTAL REFERENCES ...**193-196**

General 193
COSHH 193
Safety of electrostatic equipment and powder application 194
Standards relating to application and testing 195

Appendix II: TABLES AND FORMULAE 197

Appendix III: GLOSSARY 203

Appendix IV: FURTHER READING 209

Index: 211

Acknowledgements

We acknowledge with gratitude the following Companies and Organisations that granted us permission to reproduce diagrams and photographs;

Airflow Finishing Products Limited
Sheffield

Brittons Engineering Limited
Nottingham

Croftshaw Solvents Limited
Loughton

Elcometer Instruments Limited
Manchester

Elf Atochem UK Limited
Thatchem, Berkshire

Eurotec Surfac Coatings Systems Limited
Wigan

Hodge Clemco Limited
Sheffield

ITW Gema
Bournemouth

Morgan Newmark Limited
Telford

Nordson UK Limited
Stockport

Pollution Control Products Limited
New Chapel

Sames SA
Bournemouth

Wagner Spraytech Limited
Banbury

CHAPTER I

INTRODUCTION

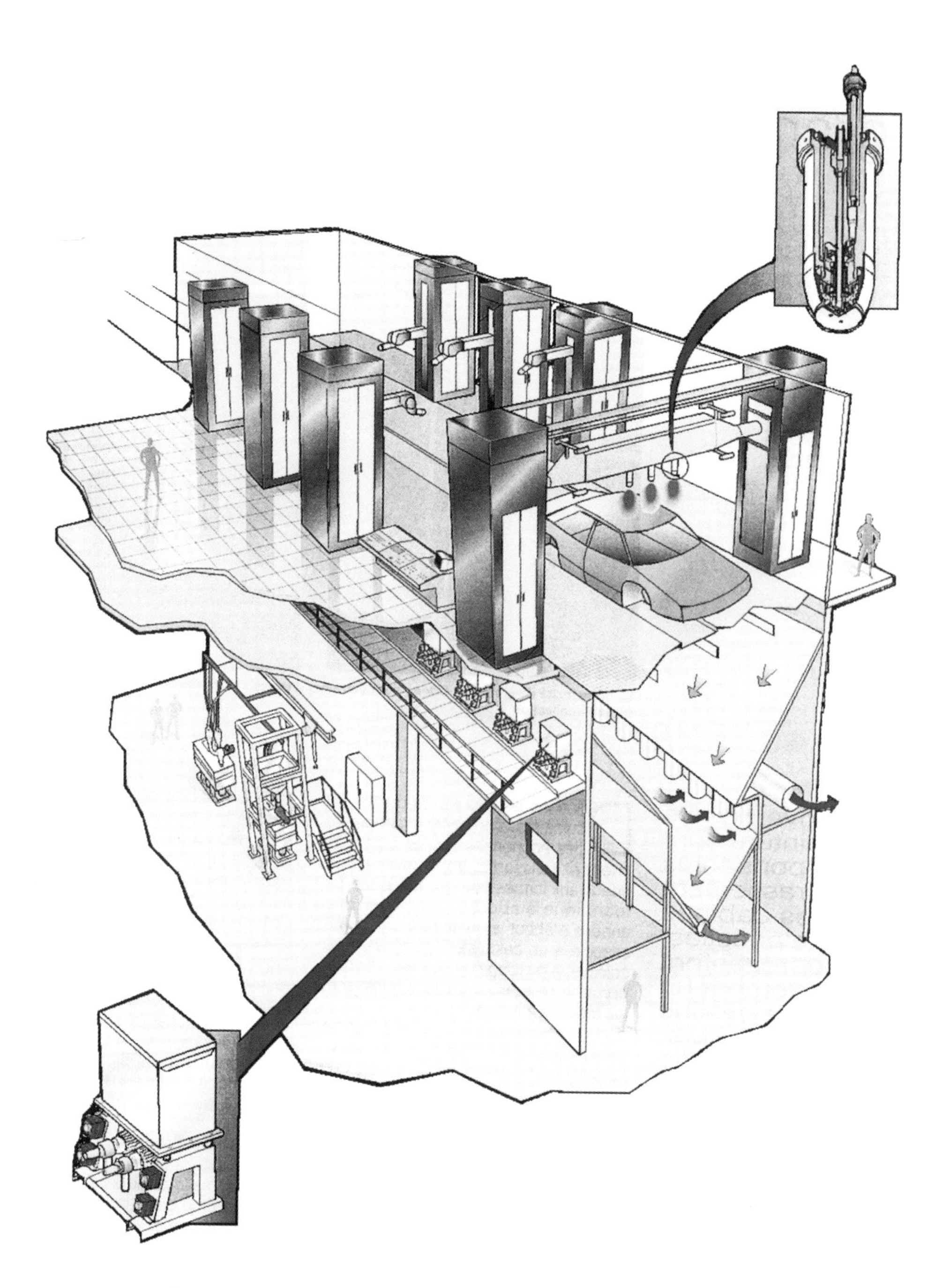

An automatic powder coating plant of the future

Chapter I

INTRODUCTION

This book describes the techniques used industrially to apply organic coating powders to an ever-increasing range of products in order to decorate and protect them.

Coating powders consist of a mixture of polymers, pigments and other components, each chosen to provide the properties required by the end user. The basic idea is that if the surface of a component can be coated with the powder and then heated, the powder is able to melt and flow out to form a continuous coating.

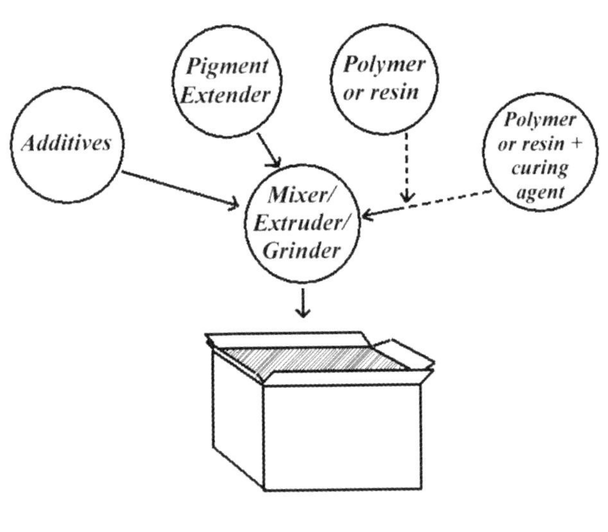

A box of powder

Powder coating is essentially a very simple process. Changes to the process are only required if customers and users ask for special properties, or if there is a conflict between productivity and the cost of the finished article.

We will be looking in detail at this simple process and will leave further refinement to be developed case by case to meet specific needs. Innovation and development of the process can radically affect the basic process, and the results have to be continuously examined in order to improve it and make powder coating a success.

As with any fairly new technology, you will probably hear lots of theories put forward about applying and using powder coatings, but these are best left to the boffins. Hands-on experience, combined with basic practical knowledge, are quite enough to meet the needs of most powder coaters.

The success of using coating powders is really a matter of keeping it simple and getting the basic principles right.

HISTORY

Powder coating has not been around for very long, the first patents being granted on the fluidised bed coating technique in the mid-1950s, so we are speaking about a technology that is barely half a century old. There had been some earlier attempts to perfect the 'flame spraying' of thermoplastic powders, and applying thermoplastics by the fluidised bed approach is still one of the simplest and easiest powder coating techniques available.

The more advanced electrostatic coating techniques are even younger and new ideas are coming along all the time. Electrostatic powder coating was first used industrially in the late 50s or early 60s.

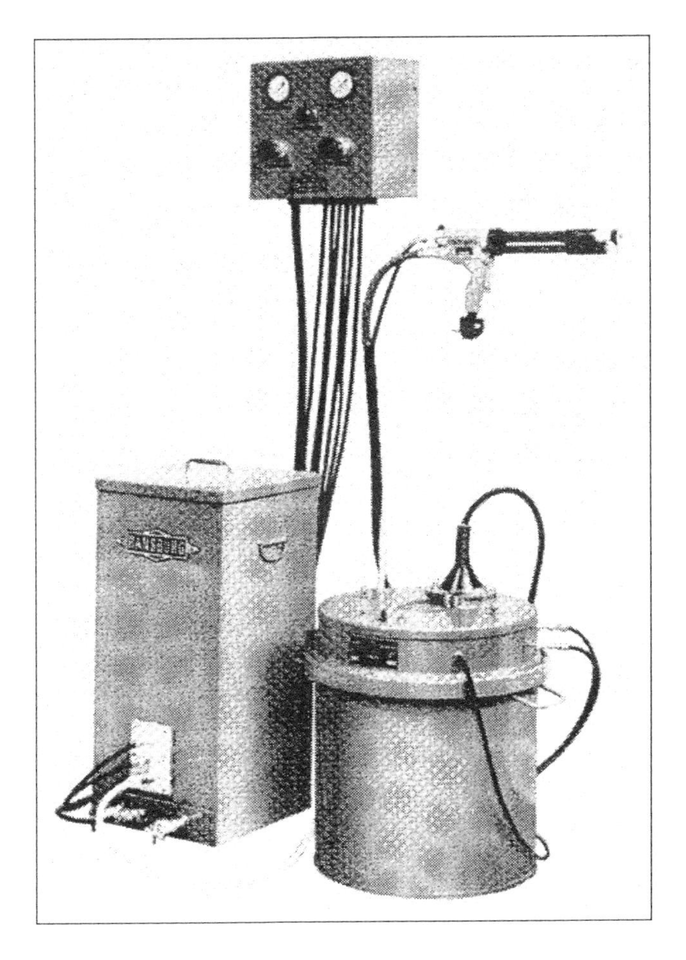

An early electrostatic unit

Looking back at the ways paint can be applied, it took a very long time to move from the brush to the spray gun and almost another 50 years to the development of the electrostatic atomiser. No one should imagine that the move to powder coating is going to be any quicker.

Coating powders will not suddenly take over and make liquid paint obsolete. Each process has its good and its bad points and coating powders and powder coatings have plenty of each. New techniques are being developed all the time and these will push both coating methods into new areas of technology in the future.

Who knows what the next fifty or hundred years will bring?

So why use coating powders?

Powder coating is usually thought of as a one-coat process, but this is less and less true as higher coating specifications are demanded.

The general advantages of powder coating are economic and environmental.

Its economic advantages are:

- energy saving
- reduced labour costs
- less wastage

The relative importance of each of these is different to different companies and depends on their own individual priorities.

From the environmental point of view, the absence of organic solvents has to be a winner. When this is coupled with reduction in waste it certainly goes a long way towards counterbalancing any potential disadvantages that powder coating may be felt to have.

In practical terms these are as follows:

- It can be difficult to change colour or coating powder type.
- The choice of colour and finish is somewhat restricted.
- Unless a multi-coat system is used it is not easy to obtain corrosion resistance at moderate cost.
- There is a limit to the minimum film weight that can be applied, and there is also some difficulty in maintaining consistent coating thickness, and matters are not helped when there are wide variations within the component itself.

- The high temperatures needed during processing can be a nuisance.

The advantages and disadvantages of powder coating will vary in importance from one user to another, for instance colour change may not be a problem to one applicator and coating thickness control may be absolutely vital to another, and so on.

THE POWDER APPLICATION PROCESS

At its simplest, the component to be coated is merely heated and dipped into a thermoplastic powder. The powder melts on to its surface, and if the component remains hot enough the powder will continue to flow out into a uniform film. Once this has happened the application process can be regarded as complete.

A great deal has been written about coating powders and their formulation. A choice of specific polymers and pigments can be made to provide a surface appearance, or finish, that will be acceptable to the end user.

Polymers are either *thermoplastic* (retaining their ability to melt when heated) or *thermosetting* (curing to a form that will not soften or melt when re-heated), but in either case it is essential for them to be able to melt and develop film-forming properties during the application process. It is clear that to produce a coating from the powder first requires the application of heat to melt the powder into a film, followed by some method of curing it if the polymer is the thermosetting type.

Looking into the future, the most critical powder coating processes are yet to be fully commercialised and will increasingly call for special powders with low curing temperature. Powders can also be envisaged that will cure by altogether different methods. Such processes are needed particularly for powder coating heat sensitive substrates such as plastics.

A coating for this purpose might for example be applied and then melted using medium-wave infrared radiation, and finally cured by ultraviolet or a similar technique, with no danger of distorting or altering the structure of the substrate.

Heat sensitive materials require ingenuity and a fresh outlook to bring these ideas together to make the solution commercially viable.

There is more to successful powder coating than may be obvious at first sight, and other processes will also have a beneficial effect on the final film.

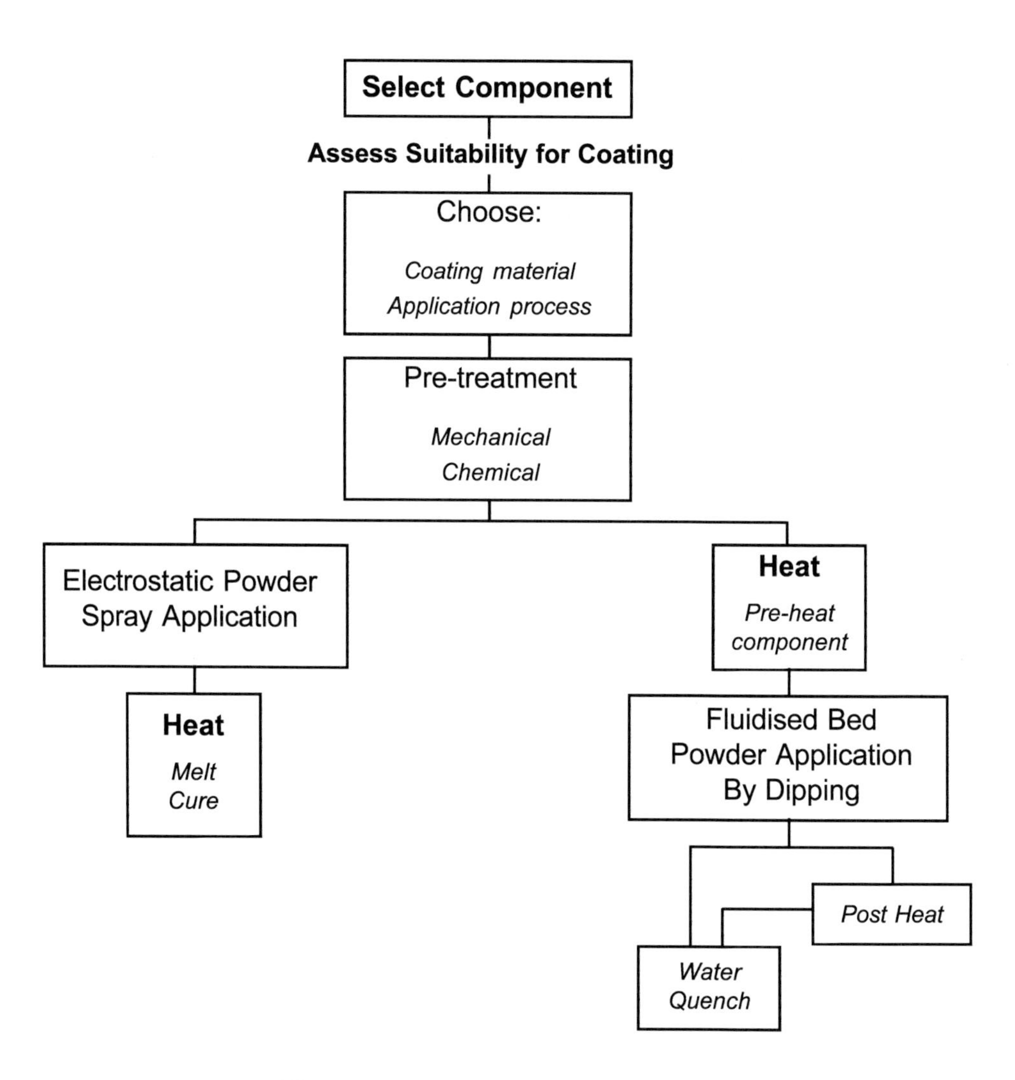

Flow chart of possible processes

The processes involved are more than just a matter of applying the coating powder. At the outset the person planning the work will need to:

- Have a careful look at the design of the component and what it is made of, and hence its suitability for coating.
- Decide on the type of coating powder that would be most suitable to give the performance required.
- Consider how the component can be held in position during coating and, if necessary, masked.
- Assess what kind of pre-treatment should be carried out before coating.
- Choose the method of application.
- Define the standard of quality required and what process controls are needed to achieve it.
- Ensure that heat required before or after coating, depending on the method used, falls within the coating powder manufacturer's specification.

The correct choice and control of all these factors will ensure that the properties of the final film will meet the end user's needs in terms of:

- Adhesion
- Corrosion resistance
- Flexibility
- Surface aspect and absence of defects, for example no contamination or obvious imperfections in the film.

Fluidised bed dipping

CHAPTER II

ASSESSMENT OF THE COMPONENT BEING COATED

Chapter II

ASSESSMENT OF THE COMPONENT BEING COATED

The design, the condition, and particularly the surface of the component to be coated, are all essential factors that need to be looked at carefully.

At the time a component is designed it should allow for the fact that a powder coating is going to be applied, either for appearance or for its protection.

The substrate must be in clean condition, free from grease or contamination, and compatible with the particular finish to be used. It also has to be able to withstand the conditions that are going to be met during processing, particularly during the heating stage.

The component should ideally be a partner for the properties of the coating powder being applied. For instance, there is not a great deal of point in applying a powder-coated film over a component that is defective or badly corroded.

Designing the component for coating

Various features of normal manufacturing practice can hinder the ability of a powder coating to give the performance required.

Sharp edges

Most coatings are thinner when applied to sharp edges and this can cause problems with both the protection offered by the powder coating and its appearance. Ideally, edges should be radiused to lessen the effect.

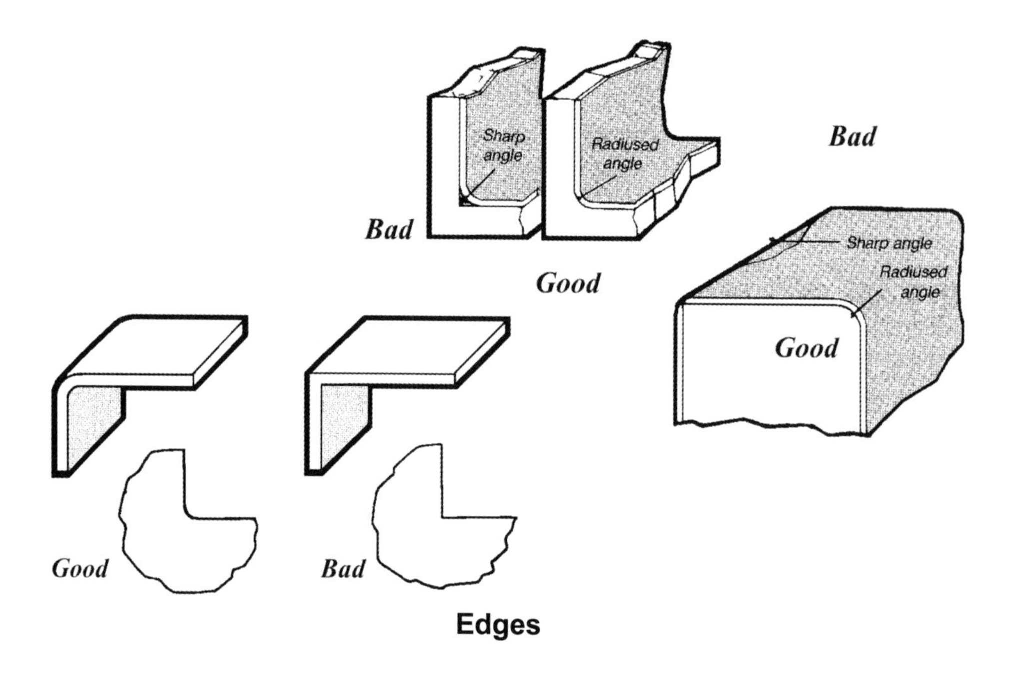

Edges

When laser cutting is used in the preparation of fabrications the problem becomes worse as the edges are particularly sharp.

Sharp edge problems occur not only on metallic substrates but also with components made of plastics, glass, and wood, and with modified materials such as MDF (medium density fibreboard).

Welds

When iron and steel components are being built up this inevitably means joining pieces of metal by welding. Welds need to be of good quality, non-porous continuous and free from splatter and oxidation. Sprays are sometimes used to assist welding, but these often contain silicone-based components and are best avoided as they are likely to cause film defects and poor adhesion. Lower temperature processes such as brazing can also cause problems as the flux used forms a hard glass-like surface that can ruin the adhesive properties of the coating. The metals used in brazing can also cause films to discolour.

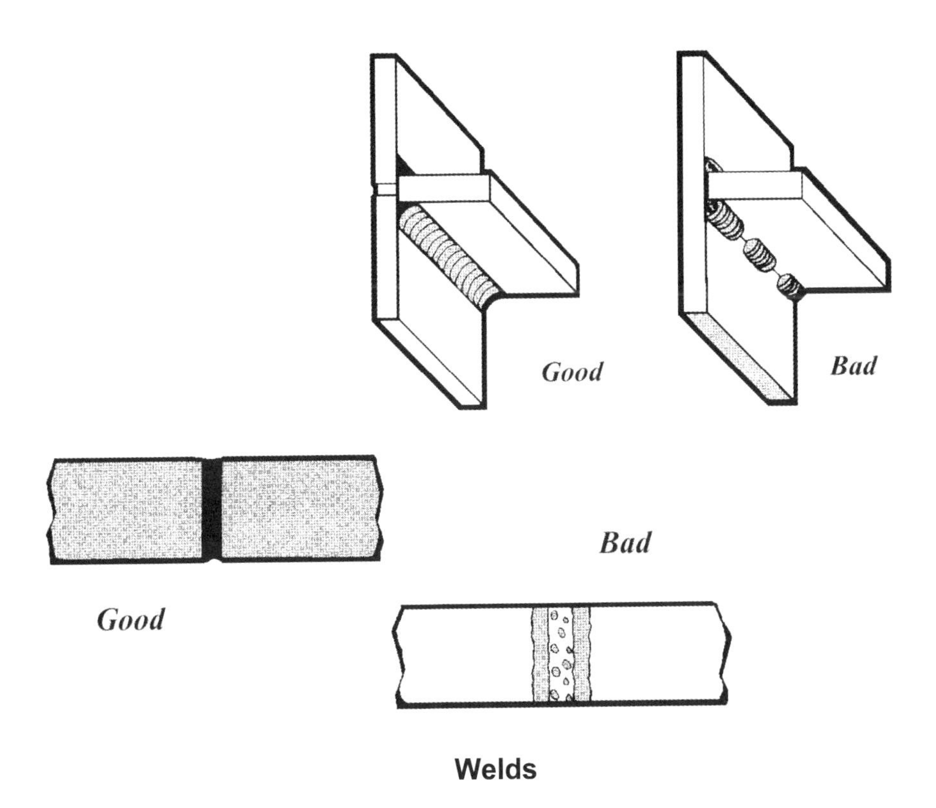

Welds

Soldered components are also unsuitable for powder coating as the melt temperature of the alloys involved is too low to withstand the processing temperatures.

Crimping

The use of crimped edges to join and finish components is acceptable providing air pockets are avoided, as these can cause bubbling during the heating process. Pockets of air can also trap chemicals used in the liquid-based processes used for pre-treatment. Sharp edges should of course be avoided.

It may be worth considering whether to carry out crimping after the coating process. Many powder coatings are able to withstand post-forming operations.

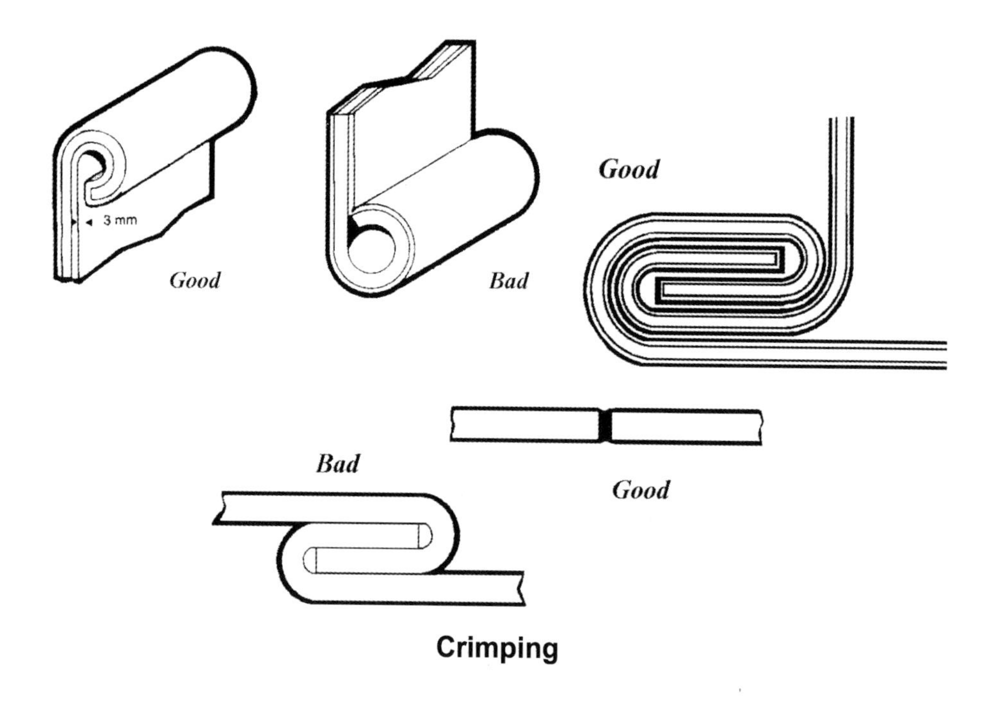

Crimping

Rolled edges

Edges of this type should not be completely closed, but any sharp edges remaining need extra protection from abrasion and damage.

Holes for bolts and screws

Allowance has to be made for any coating that will be applied later to internal bores – or in fact for the bolts and screws themselves if they are also going to be coated.

Threads

In general it is not practicable to coat threads, unless they are particularly large and the coating thickness has been carefully allowed for. The coating of threads is a specialist operation.

Components with moving parts

Assembled components with moving parts such as hinges and fastenings do not coat well as the coating film will bond the moving parts together. Once the bond is broken in use it will leave the substrate uncoated at this point.

In this case the components must be coated before assembly and the coating thickness allowed for at the design stage. The coating can however often provide an excellent non-stick bearing surface as well as offering corrosion protection in enclosed areas.

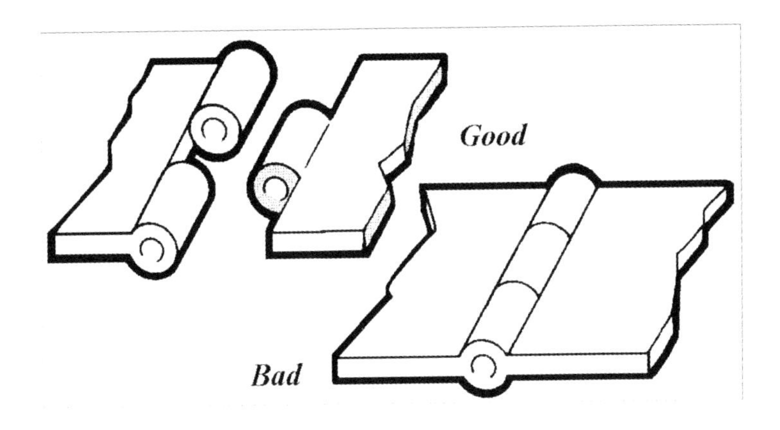

Hinges

Allowing for coating thickness

A characteristic feature of powder coating is the thickness of the coating that can be produced, often in excess of 100 microns, depending on the technique used.

When the coating is particularly heavy, for example when multi-coat systems are applied in order to achieve specific film properties, problems can be encountered when the coated components are finally assembled if the coating thickness has not been allowed for at the design stage.

At all stages of the process the unusually high thickness of the coating has to be borne in mind and planned for.

THE SUBSTRATE

The final performance of a powder coating is only as good as the condition of the substrate it is applied to. The substrate should always be inspected to look for potential problems so that steps can be taken to avoid them. Once applied, faulty powder coatings are difficult and expensive to put right.

A wide range of substrates and components are suitable for powder coating. Until recently they have tended to be either ferrous metals such as mild steel or non-ferrous materials like aluminium, zinc die-castings and alloys. As the technology improves, however, other substrates are increasingly powder coated, including glass, many types of plastics, some woods, and particularly composites such as MDF (medium density fibreboard).

It is essential in all cases for the substrate to be clean and free from contamination. If metals are involved they should be non-porous and free from corrosion and scale.

Powder coatings applied to substrates showing any of these problems will give poor adhesion, bubbling, pin-holing, and a number of other surface defects.

It is often said, and it is probably not an exaggeration, that 90% of a coating's properties come from the pre-treatment, and that 90% of the effectiveness of the pre-treatment depends on how well the initial degreasing was done.

Proper pre-treatment of the substrate before coating is therefore vitally important. The factors that will ensure that the condition of the substrate or component is satisfactory when the coating is applied need to be carefully thought out and controlled.

Each substrate presents its own problems to the powder coater.

Metals

The application of coating powders to metallic substrates presents different problems depending on which metal is involved. In all cases contamination, either in the form of corrosion or other materials on the surface, are always a serious threat to the success of powder coating.

Ferrous metals

Surface contamination such as mill scale, corrosion, oils and grease must of course be removed. There are great differences in grades of steel and each provides its own problems.

Cold-rolled, pickled and oiled steels are relatively simple to pre-treat. Efficient degreasing will be the initial treatment. Hot-rolled steel can be a poor surface for coating without some vigorous chemical or mechanical action to bring about a change in its condition.

The use of lasers for cutting profiles can upset the adhesion of the coating unless the gases used to shroud the laser are chosen with care. Some gases can produce a glass-like surface that will not give good coating adhesion. This will have to be roughened to remove the oxidised surface layer.

Non-ferrous materials

Alloys containing aluminium and zinc need thinking about carefully. The quality of castings can vary considerably and porosity is a particular problem.

Extruded materials can also be a challenge if the highest levels of adhesion are required, and the pre-treatment processes have again to be chosen with care. Extrusions need to be handled gently, as their ultra-smooth surfaces tend to 'bruise' and cause blemishes that are visible in the final film.

As usual, degreasing is essential but unlike ferrous metals severe chemical or mechanical action is not a good idea as it can cause damage to the surface by pitting or corrosion. 'Smutting' may appear and an additional process is then required to get rid of it.

Galvanised components

The surface of hot-dipped galvanised components can be surprisingly variable. Oxide and slag is often deposited during the galvanising process, and galvanisers will sometimes 'passivate' the surface after dipping.

There is no simple pre-treatment for galvanised surfaces. Each batch of work needs to be inspected and assessed, and a decision made as to whether chemical or mechanical pre-treatment is advisable before powder coating.

Many applicators find it useful to apply a liquid self-etch primer or other acid-activated material. Gentle grit blasting can also help by providing a uniform bright surface ideal for powder coating.

Zinc-plated ferrous components

These are best treated in the same way as non-ferrous components as far as chemical pre-treatment is concerned, but care needs to be taken not to damage the plated surface.

Glass

Glass components are often powder coated without pre-treatment, but great care must be taken when handling them not to introduce contaminants, especially finger marks.

Plastics

Powder coating is increasingly used for plastic components. The plastic must be able to withstand the temperatures used during both the coating itself and the curing of the coating, without distortion or degradation occurring.

There are many opportunities in this area, particularly where solvent-based systems have been used in the past, and exciting new techniques for this application are being seen every day.

After application the coating powder is generally melted using a controlled infrared source to provide flow-out and the coating is then ultraviolet-cured without additional heat.

Plastics do not usually need pre-treatment. As in the case of glass, careful handling is important as surface contaminants have to be avoided, once again including finger marks.

Wood

This is a relatively new area for powder coating.

As usual, the surface of the wood must be contamination-free. Its moisture content also needs to be carefully controlled within specified limits. Electrostatic attraction of the powder is improved with increasing moisture content, and this can of course be an advantage, but if it is too high it will cause defects in the film. The satisfactory target range is 3 – 4%.

With wood-based materials of low moisture content it is often a good idea to pre-heat the surface to help coating, particularly in the case of MDF. A controlled infrared source is ideal for this purpose.

Curing has to be carried out carefully in order to protect the base material from degradation and to allow the ‘out-gassing’ of air and moisture through the film.

The choice of coating powder type is also an important factor in the successful coating of wood-based substrates.

ADHESION

Powder coatings are applied to a variety of substrates or components for decorative, protective, and functional purposes. In each case it is imperative that the coating adheres well to the substrate. This may seem obvious and straightforward but coating adhesion is in fact a complex subject.

The applicator must always aim for the highest possible adhesion in order to allow for unknown or unforeseen conditions that a component may come up against in use. It is very important that coating failure should not occur either in the short or the long term.

To obtain the maximum performance from a powder coating the adhesion of the film to the substrate is therefore top priority. Almost all of the preparation processes mentioned so far are there to help adhesion. It is useful for us at this point to look at the basic principles involved.

The main causes of failure are likely to be:

- Surface effects.
- Chemical reactions at the surface.
- Surface irregularities.
- Contaminants such as oxides, oil, dust, absorbed water, etc.

A number of techniques have been developed with the aim of modifying the surface being coated to help adhesion, and we have already seen that it is important to choose the correct pre-treatment process for the type of component concerned.

Coating powders are complex formulations, and the chemistry of the polymer is an important factor in promoting adhesion on a given type of substrate.

Weathering failures such as blistering and scab corrosion are usually regarded as problems of adhesion. Weathering can sometimes cause coatings to delaminate (peel off), even when the substrate itself has not corroded.

A common problem involves the interfacial chalking of epoxy-based coatings. In the presence of sunlight, moisture and oxygen the epoxy powder can become degraded at the point where it makes contact with the surface of the component, and this causes delamination.

It is obvious that it is essential to match the coating to the substrate and to ensure that the pre-treatment is correct. After this, the application and the curing of the coating must both be carried out properly if good adhesion is to be obtained.

Adhesion is the result of the attractive forces between the coating and the substrate that combine to hold them together. We will now have a quick look at adhesion and some of its simple theoretical background.

Wetting contact theory

The wetting contact theory arises from the observation that the physical forces involved in adhesion between coating and substrate are mainly a matter of how well the surfaces are able to 'wet' one another. The importance of wetting on good adhesion follows from the fact that the greater the area and the degree of contact, the better the adhesion. When coating powder is applied it must therefore be able to wet the surface effectively.

The larger the surface area the better the adhesion will be, and a rough surface should therefore in principle provide a better grip than a smooth one. If it is too rough, however, the coating will not flow out well enough on melting to cover the roughness, and this will show in the final result.

The effect of surface contamination through poor pre-treatment – or no treatment at all – is to cause reduced wetting, and this is the first area to be looked at whenever an adhesion problem crops up.

The coating powder will have been formulated so that in ideal conditions it is able to wet the surface properly when it melts and flows out. During this process the time and the temperature are key factors in the case of thermosetting materials, as the powder must have time to melt and flow out before the coating begins to cure (or 'set') by polymerisation. Good adhesion is clearly not going to result unless enough time is allowed for the coating to completely wet the surface before this happens.

In this respect it is easy to see that solvent-based systems have an advantage over powder coating since the solvent is likely to be able to wet the substrate more easily, and it may in addition be able to dissolve away contaminants on the surface.

Mechanical adhesion

This is a special case of the wetting process described above. It is found in practice that there is an extra bonus if the surface is cleaned and roughened using blasting techniques, thought to be due to the mechanical locking of the coating on to the peaks and troughs formed on the substrate. If however there is not also good wetting between the coating and the substrate then the increased roughness produced by blasting can actually lead to a *decrease* in adhesion by producing uncoated voids at the interface.

It can be difficult to get coatings to stick satisfactorily on to some types of plastic substrates and chemical etches are used to form cavities or craters in the surface, giving the coating a better chance to key on to it by mechanical means.

On the other hand it is interesting to find that excellent adhesion can be obtained on some completely smooth surfaces such as glass, so it is not always necessary for mechanical adhesion to play a part.

Chemical adhesion

Chemical bonds are much stronger than the physical forces described, and coatings that have the ability to react chemically with the surface of the substrate will therefore have particularly good adhesion. There are numerous applications where chemical interaction is useful for promoting adhesion.

Dramatic effects can be obtained, for example, using isocyanates and epoxies, each of which can be formulated to react with the surface of the substrate and form bonds that are very difficult to break.

HANDLING OF COMPONENTS

It has been mentioned already that the handling of components being powder coated is important. This is easy to understand in the case of fragile materials like glass, and soft items such as plastics, wood and non-ferrous metals that can easily be bruised or dented if handled roughly.

Ideally, glass and plastic components should be taken from the pre-treatment process without touching with the bare hands, for example using lint-free gloves. Items should be packaged before taking to the coating area to ensure that they remain free of dust, moisture and other forms of contamination.

Devices for holding components during coating, such as hooks, hangers and other jigging systems, must be stable, strong enough and easy to use. Handling equipment can be custom-made for particular components but it is becoming cheaper and more convenient to buy standard hooks, masking devices and other jigging utilities in bulk.

During pre-treatment

Components need to be held securely during the cleaning operation. When they are to be treated mechanically it is often necessary to protect certain areas and this can be incorporated in the design of the handling device.

When chemical processes are used other factors become important:

- Protection of those parts which are not going to be coated.
- Ensuring good wetting of all areas by the chemicals used.
- Hanging the components correctly to allow proper drainage and to avoid chemicals becoming trapped.

After pre-treatment

Once a component has been prepared, free from contamination and often with a conversion or etch coating applied, it is a good idea to apply the powder coating as soon as possible. During any delay components will tend to collect moisture, dust, and other contaminants very quickly and all the good work achieved so far will be undone.

The components must once again be treated with care and should *never* be touched with the bare hands. Clean lint-free cotton gloves are used, as grease and finger marks will be visible in the finished powder-coated article.

If a delay between pre-treatment and coating is unavoidable, a storage area and suitable packaging should be provided to keep the components clean, dry and absolutely free of dust.

During application

The spray coating process often requires the component to be movable, or if this cannot be done then it must be possible to move the spray gun instead. The jigging of components is important and needs to be planned so that the coating can be applied correctly and economically.

During coating the operatives need to be able to see what is happening, particularly in critical areas, both in fluidised bed coating and in electrostatic spraying. The type of jigging device used may be completely different for dipping and spraying, but the objective is the same.

Dipping requires the components to be rigidly mounted so that they can easily be manipulated while they are dipped, rotated and perhaps blown with low-pressure compressed air to remove excess powder. Water quenching may follow this.

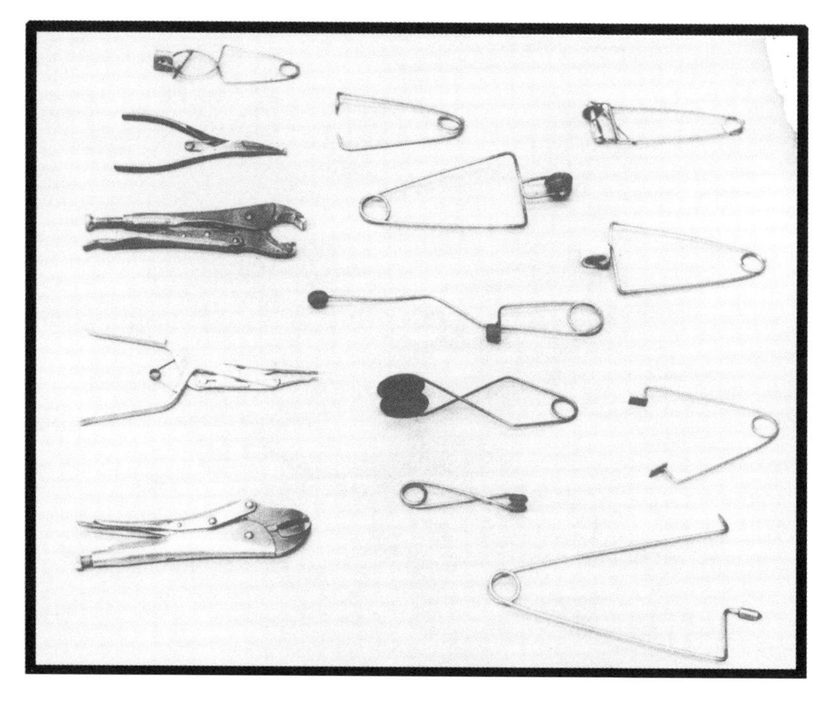

Proprietary hooks and fixtures for dipping

The electrostatic spraying process requires a means of hanging the components that avoids masking or shadowing in any of the areas to be coated, and which also makes minimum contact with them.

Good earthing is required to ensure that static charge is dissipated and does not build up, inhibiting transfer of the powder during the application process.

To summarise, the important points to be borne in mind in the design of jigging and hooking devices are as follows:

- Strength.
- Stability.
- Earthing – in the case of electrostatic application.
- Drainage – in chemical pre-treatment and fluidised bed dipping.
- Minimum shielding during coating.
- Suitability for re-use after stripping.
- Cost.

Conveyor loading

If components are being carried into the coating area by mechanical conveyor, a number of additional factors become important:

- The minimum distance possible between components without shielding.
- How much weight can safely be carried.
- Space to be allowed for colour changes.
- Allowance for rise and fall in the conveyor.

Cleaning of hooks and jigs

The usual methods for cleaning ('stripping') the hanging devices are:

- Controlled incineration.
- Heating and degradation – fluidised bed cleaning.
- Chemical cleaning.
- Blasting.

Which of these is chosen depends on how well the device is able to withstand the treatment.

Incineration or pyrolysis

Controlled pyrolysis is used for heavy-gauge components able to withstand temperatures in the range 550 - 800°C for up to 8 hours. The items to be stripped are often sprayed with water during heating to control the degradation of the coating. The products of combustion are passed though an incineration chamber at even higher temperature to burn off potentially toxic materials before venting to atmosphere.

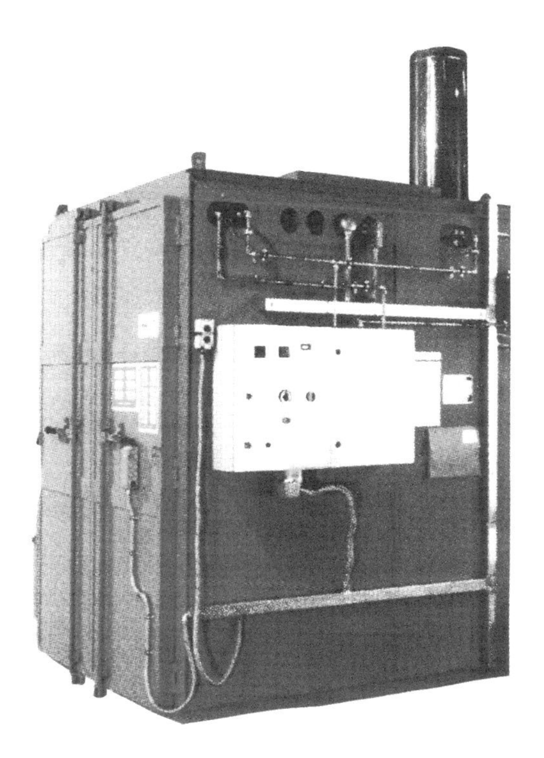

Controlled pyrolysis oven

After pyrolysis, ash is removed from the jigs and other items and after cleaning to ensure they are dust-free they can be used again.

Fluidised bed cleaning

Stripping can also be carried out by heating in a high-temperature fluidised bed. During this process the coating becomes degraded and brittle and finally falls off.

Chemical stripping

Treating the coating with a paint-stripping solvent such as methylene chloride can be very useful for cleaning equipment. This material is toxic and it must be handled with due caution. Components need to be washed off afterwards to remove all traces of the solvent.

High-boiling solvents and chemicals such as methyl pyrrolidone are also used. These work by softening the coating material and delaminating it from the substrate. The process is carried out at 50°C, with regular additions of sodium or potassium hydroxide to assist the stripping action. The solvent is kept in reasonable state by removing solid matter by filtration but the stripped component still requires to be cleaned off afterwards.

The tank of solvent will deteriorate as it becomes more and more contaminated. This makes the process costly, as spent solvent requires either to be recovered or carefully disposed of by specialist contractors.

The health and safety and environmental issues need to be assessed for this process, and emergency procedures must be in place in case of an incident.

Mechanical stripping

Abrasive blasting using metallic or plastic particles will remove the coating slowly but effectively. The components must obviously be strong enough to withstand the impact involved.

The blasting media will in time become contaminated and must be sieved and separated to remain effective. Contamination of the surface of the handling devices from such residues must at all costs be avoided. Any material of this kind on a hook or jig will degrade during any heating process and can then find its way on to the component, causing defects in the coating film.

Masking

It is often necessary to avoid coating being applied to certain areas of the component. This can be achieved by the use of a variety of masking materials and devices.

High temperature adhesive tapes are convenient and are often used. They are very effective but need to be removed before the component becomes cold as otherwise a portion of the adhesive tends to remain behind.

A variety of caps, profiles, and inserts made of high-temperature rubber or PTFE are also available for this purpose.

Intelligent use of masking can certainly broaden the scope of powder coating. It is worth remembering that it is much easier to mask articles being powder coated than if conventional liquid paint is used.

Masks used in electrostatic powder application can be removed before the coating has been melted and cured. In this case they can be taken into a cleaning area to blow off dry powder and re-used straight away.

Masking

CHAPTER III

PRE-TREATMENT

Chapter III

PRE-TREATMENT

The processes used for pre-treating articles before powder coating are numerous, often complex, and largely proprietary. A great deal of care needs to be employed when selecting which one to use.

A practical test is often the best way to approach the problem. A component may be pre-treated in a variety of different ways, then coated and the overall performance of the coating evaluated.

The most useful assessment is an adhesion test, followed by other destructive tests such as salt spray and humidity evaluation.

In many cases the suppliers of the raw materials will help by recommending the pre-treatment process and it is important to follow their instructions in detail.

The processes available are as follows:

- Simple wiping – removal of dust and other contamination.
- Degreasing – using solvent or water-based materials.
- Degreasing – using solvent vapour systems.
- Mechanical treatments – abrading using hand tools and other equipment.
- Mechanical treatments – by grit blasting.
- Water washing with chemicals at high pressure and temperature.

- Chemical treatments – for ferrous and non-ferrous materials, respectively.
- Other chemical treatments.

WHY PRE-TREAT?

Once the surface of the substrate has been assessed for its suitability for coating, it has to be prepared so that the maximum degree of protection can be achieved. Any contamination on the surface should be inspected and investigated. Is it present on every component or is it just an occasional occurrence? Is there any corrosion? Whichever process is chosen must take care of all these eventualities.

Good preparation means identifying potential problems and their causes, and choosing the corrective action required:

i) Removing contamination
 – grease and oils
 – corrosion
 – dust or other particles

ii) Modifying the surface
 – abrasion
 – chemically

iii) Coating the surface to enhance adhesion and corrosion resistance
 – chemical conversion coatings

Removing contamination

Thorough cleaning is the single most important factor in obtaining an acceptable coating. Contaminants may be petroleum-based oils, greases, polymers, or surfactants such as soaps.

Greases and oils will upset the wettability of the substrate and prevent the coating powder flowing along the surface after melting. In addition, the contaminants may dissolve into the final film and cause pinholing and discoloration.

Soluble mineral oils and water-based emulsions are simplest to remove, but some soapy materials such as stearates can be very difficult. Simple testing before full production will establish the most suitable method.

All corrosion must be removed from ferrous and non-ferrous metals as it will affect the ability of the molten coating powder to wet the surface. Unless it is thoroughly removed corrosion can also continue to spread under the final film. This results in loss of adhesion and as it creeps slowly along the substrate it will finally break through the coating.

Red rust and white non-ferrous oxides are best removed using acidic solutions at the degreasing stage.

It is similarly important to remove all dust and other particles from the surface. During coating, contaminants of this kind can travel up through the melting film and cause surface defects and blemishes.

Glass and plastic components are particularly liable to this type of problem since the surface of the component soon becomes charged with static electricity, caused by handling and movement in the packaging. Dust then becomes readily attracted. The dust itself often has a natural static charge and this makes the problem even worse.

Dust is best removed using either acidic or alkaline solutions, possibly adding a surfactant; the presence of the surfactant can help to prevent further build-up of static. Only by practical testing can the ideal treatment and method be chosen.

Components made of wood or MDF (medium density fibreboard) will often have been sanded after cutting and shaping, and any dust remaining from this must be thoroughly removed. Chemical treatment is not suitable on wood, so mechanical methods have to be used such as brushing or blowing off the dust with de-ionised air blowguns.

Mechanical modification of the surface by abrasion can improve the adhesion of coatings to metals, plastics and glass. This is carried out by blasting with a variety of materials to change the nature of the surface.

Surface modification can also be produced chemically and this can give dramatically better adhesion. Unfortunately some of the chemicals used for the purpose are rather unpleasant.

Alternatively, applying a *conversion coating* enhances both adhesion and corrosion resistance in one go, and this is probably the commonest process used on metallic substrates before powder coating.

Conversion coatings lay down the chosen chemical in a fine crystalline structure at the interface and provide an excellent surface for coating. There is a wide choice of materials available:

- Iron phosphate
- Zinc phosphate
- Chromium phosphate
- Chromium oxide

Other materials are at present being actively considered as replacements for those containing chromium, as these are toxic.

Molybdates, silanes and other similar materials are also commercially available and have similar or sometimes better properties in conversion coatings.

Methods of pre-treatment

Wipe cleaning

This rough and ready process should not in itself be seen as an effective pre-treatment for powder coating but merely as the starting point for other methods.

In most cases, the other pre-treatment techniques used should not require components to be pre-wiped before spraying unless they have become dusty in storage. If it is needed, special purpose lint-free cloths should be used, taking care to avoid snagging on sharp edges and creating further dust and debris.

Wiping with large amounts of volatile organic solvents is not desirable on environmental and safety grounds unless specifically approved for the purpose. The use of solvents for local spot cleaning has the advantage of minimising this problem as it is only used over small areas of large items, for example when finishing off a component already cleaned and pre-treated.

Solvent dispenser for wiping using special cloths

Water-based and other systems of low volatility are more and more in use nowadays, but the time it takes for the component to become completely dry can be a big disadvantage.

There is a tendency for wiping to produce more waste than other procedures. The cloth becomes loaded with both cleaning chemicals and contaminants and has to be changed frequently. This method can also have health and safety and fire risks, and special precautions are needed with many of the solvents used.

If it is found that repetitive or large-scale wiping is tending to become standard practice, the procedure should be reviewed to see whether an alternative method might be better. Dusting off with de-ionised air blowguns could be worth thinking about.

MECHANICAL PRE-TREATMENT

Abrasive cleaning is used in the following circumstances:

- To remove oil, swarf, grease and corrosion from the surface of a metal.
- To remove dust and other contaminants from non-metallic substrates.
- To smooth off laser-cut edges.
- To provide a surface for good coating adhesion.

Spent abrasive tends to be an expensive item of waste in this type of treatment. On the other hand inadequate cleaning leads to reworking, which wastes both time and materials. A variety of mechanical techniques are available, ranging from simple brushing to shot blasting.

Brushing

This is an effective method for removing scale, rust, previous coatings, and other tightly adhering contaminants. It is not suitable for removing fluids, however. Where surfaces are heavily contaminated with oils and grease, these will stick to the brush and spread the contamination rather than removing it.

The quality and cleanliness of the brushes govern the final finish. They should be regularly knocked, scraped or washed to remove build-up.

Abrasive pads

These can be used to clean, roughen or scuff the surface, either manually or using powered hand-tools. Bonded or woven plastic pads produce an excellent surface for promoting adhesion on many surfaces before applying the coating.

Shot or grit blasting

This method uses compressed air to fire abrasive particles at the surface of the component, removing contamination, corrosion and weld residues, and leaves an ideal surface for good adhesion.

Grit blasting

Blasting can employ a broad range of abrasive materials:

- metal particles

    – chilled iron shot
    – steel grit
    – aluminium oxide

- plastic pellets
- glass beads
- natural materials such as crushed nut shells
- solid carbon dioxide.

The term *sand blasting* is often used to describe this process, but sand itself, as well as other materials containing free silica, are not permitted for blasting purposes for health and safety reasons.

Given the increasing cost of disposing of solid waste, it is worth having a careful look at the way you operate the blasting process to make quite sure that the blasting medium is high enough in quality and is being used effectively. Modern equipment allows the blasting material to be collected and permits it to be recovered by separating it from contaminants removed from the substrate. Unless correctly maintained and provided with a good filtration system, grit blasting can be a constant source of contamination.

It is best for components to be degreased before shot blasting rather than after. Once the abrasive becomes loaded with oil and grease it is increasingly difficult to obtain a sound clean finish; if it continues in use, these contaminants are merely transferred back on to the surface.

It is important to control the nature of the finish obtained. The force of the impact used, and the size, density, hardness and shape of the shot or grit will dictate the degree of cleanliness and the nature of the surface produced.

The surface condition of mild steel, either before or after blasting, is often defined using Swedish Standard SIS 055900:

Grade of finish	Description
Sa O	No treatment
Sa 1	Light blasting to remove rust and loose mill scale - slight brownish tinge
Sa 2	Vigorous blasting to remove mill scale and corrosion - grey finish with slight discoloration
Sa 2.5	Discoloration removed
Sa 3	Cleaned to white metal

British Standard BS7079:1989 is similar.

Surface profile is the term used to describe the microscopic character of the surface. A smooth profile will not give as much adhesion as much as one that is rougher, but at some point the coarseness of the finish will start to show through in the appearance of the final film. A profile with around 25 microns between peaks and troughs is about right.

Grit blasting is a severe process and can even deform ferrous substrates. Components manufactured from non-ferrous metals, glass and some types of plastics are even more easily damaged.

Blasting processes can be carried out either manually or using automated equipment. It is important to remove dust afterwards in an area well away from the coating operation. In all cases operators must be protected from noise and dust.

Chemical cleaning

Cleaning operations, particularly those involving chemicals, are expensive and it is worth asking oneself to start with whether cleaning is really necessary. In some cases it can in fact be avoided.

In any case it is a good thing to try to reduce the amount of contamination arising in the first place, for instance by improving handling procedures and by removing some soiling in advance by mechanical means. Chemical cleaning will then be that much easier, reducing the amount of solvent needed and prolonging the life of the cleaning bath.

We have seen that the powder coating of metals requires the substrate to be free from oil, grease, swarf and other contaminants, including any previous coating. To be on the safe side it is advisable to treat glass and plastics in the same way.

The coating of plastic articles can be simplified if mould-release lubricants have either not been used or have been chosen with care. If plastic components are removed from the moulding process using gloved hands and carefully packaged for transportation to the coating operation, it may only be necessary to remove dust before starting coating. A similar clean handling regime can equally be used for glass components.

Some form of chemical cleaning is normally required, however, in cases where mechanical treatment is not quite good enough to do the job properly. If both are needed, the chemical cleaning should be carried out first in order to prevent contamination of the materials used in the mechanical process.

A wide range of cleaning agents and methods are available. When considering which to use, the following should be borne in mind:

- The suitability of the process for the component and the type of contamination present.

- The thoroughness of the process in relation to its cost.

- How well it will affect coating adhesion.
- Its effect on the coating surface – whether defects will recur, etc.
- Health and safety considerations.
- Its environmental impact.

As an example, a cheap substitute material may be first-class for cleaning a surface, but it may bring with it a fire or health risk, or require consent for the rinsing water discharged to sewer.

Many standard cleaning methods using traditional solvents can now be adapted to alternative cleaning agents that have no environmental and health concerns.

Vapour degreasing

This is the most effective way to degrease many types of substrate. It can be hazardous and environmentally unfriendly, but using modern control techniques the hazards are minimised and environmental legislation can effectively be met.

When an oily or greasy component is placed cold above a boiling solvent such as trichloroethylene, the vapour condenses on the cold surface of the component and dissolves any soluble contaminants. The liquid then drains back into the hot liquid below and cleaning continues as further vapour condenses on the component. Once the temperature of the component reaches that of the vapour, condensation stops and the cleaning process is finished. The component is then withdrawn slowly and allowed to cool to room temperature again.

Some solid and other non-soluble material is also washed away during this process. Complete removal usually requires immersion in the boiling liquid, or an alternative technique such as spraying or the use of ultrasonics. Cup-like recesses on a component require special attention.

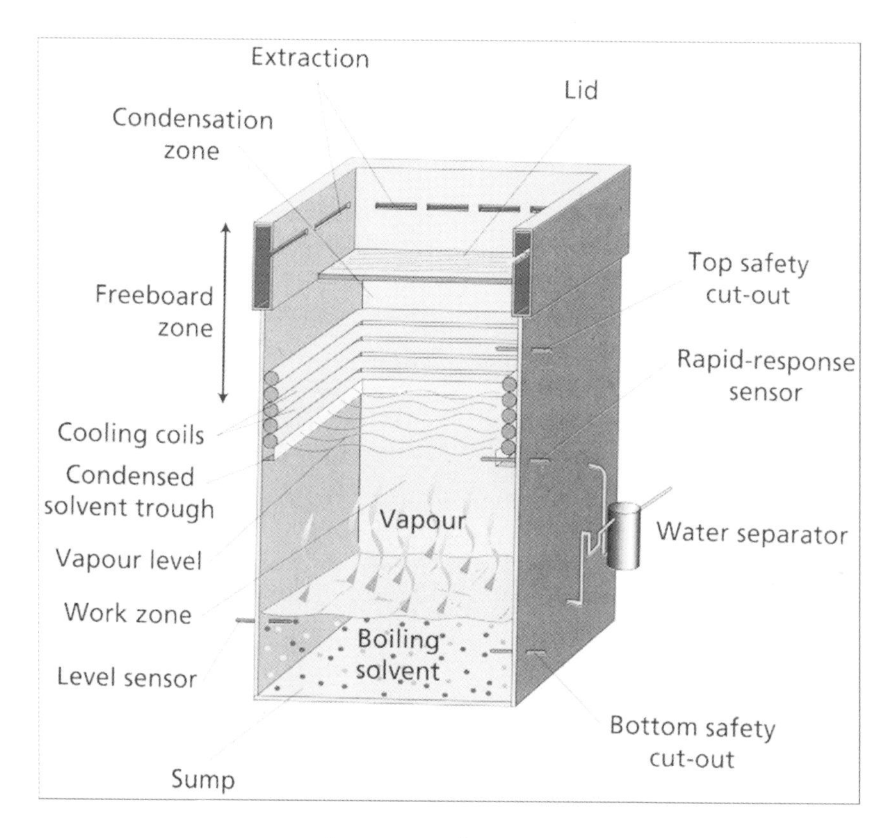

A vapour degreaser

Vapour degreasing is most effective with components of high heat capacity. It follows that thinner sections are not cleaned as effectively because they heat up more quickly and reduce the amount of solvent condensing.

Several types of solvent are used in this process, trichloroethylene being the most popular. Brominated hydrocarbons are also used. All are to some degree volatile and will need to be replenished from time to time due to evaporation. Discharge into the atmosphere has to be minimised since:

- The vapour can affect the environment, for example by ozone depletion.
- It can be harmful to animal life, including humans.
- Solvent loss has to be made up and makes the process less economic.

Conveyorised tunnel systems

This method is similar, but the degreasing agent is applied within a confined space. Vapour immersion times and withdrawal rates are planned and controlled. Additionally, the degreasing materials can be automatically sprayed on to components suspended on jigs or flight bars.

Careful design of the plant and extraction system is needed to avoid solvent losses and to keep within the law, but this technique is cost-effective for medium and high volume coating plants.

A conveyorised degreaser

Ultrasonic cleaning

The use of ultrasonic energy is effective for removing many types of contaminanat cheaply and effectively. All areas of the submerged parts are cleaned, and sonic energy is in most cases better at dislodging solid residues that vapour or static degreasing systems.

Initial trials should be performed with certain substrates; for example, erosion can be observed on some metal surfaces.

Either aqueous or solvent systems can be used, and it is often found possible to adapt existing tank systems to this procedure.

Spray cleaning

This is a method applicable to many components and contaminants and is particularly good for removing particles of debris, providing the spray is directed to all the areas concerned. The operation is usually performed in a glove box or other small enclosed space.

Water-based materials are normally preferred; flammable liquids should be used only when proper safety precautions such as fire suppression systems and inert gas blanketing are in place.

The addition of detergents with low foaming characteristics is recommended for both high and low pressure spraying. Hot-water sprays containing surfactant are useful for cleaning large components in cases where it would be difficult to immerse them in a tank.

Power-wash cleaning

This water-based process combines several techniques such as immersion, spraying and sometimes also ultrasonics in a single automated machine. This needs less floor space than an equivalent in-line system.

The system can be thought of as rather like an automatic dishwasher as washing, rinsing and drying cycles are all carried out in one tank. The machine can also incorporate a custom-spray device to clean blind holes. It has the advantage of good material control and low labour costs.

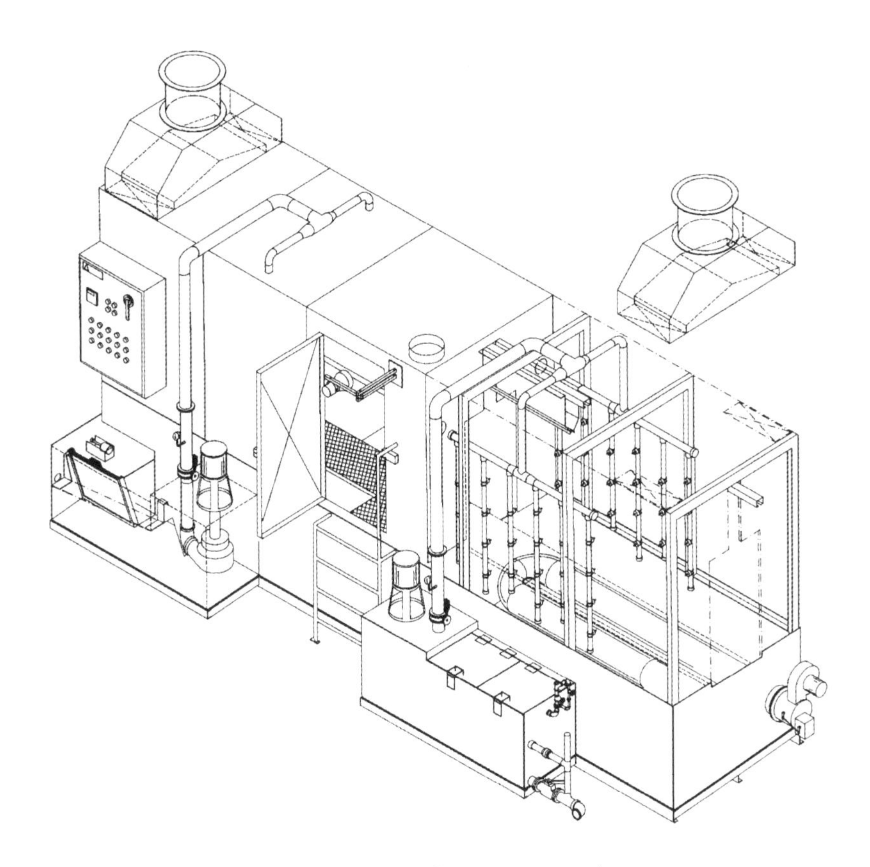

A self-contained power washer

Steam cleaning

Water-soluble contaminants, as well as oils and grease, can be removed using portable steam-generating equipment. This requires little floor space and is useful for the infrequent cleaning of large objects as it can be carried out manually. Additives such as alkaline detergents and rust inhibitors are employed.

Care has to be taken not to leave detergents on the cleaned substrate as they can later cause poor adhesion of the powder coating, and a final rinse is therefore desirable. With some metals an additive is included to produce a very light layer of iron phosphate, which prevents flash corrosion.

In all these operations the local water authority has to be involved if the effluent, accompanied by the contamination removed, is to be discharged direct to the main sewer. It may perhaps be necessary to install a settlement tank to remove solid matter.

Alternative cleaning materials

A wide range of alternatives to traditional organic solvents are now available. Many of these are still organic compounds but are better because they are less volatile, and some have the added advantage of increased cleaning power combined with relatively low risk to human health and the environment. Examples are low volatility esters and alcohols.

Water-based systems are better still from the environmental point of view but the low evaporation rate of water sometimes makes substitution into existing processes difficult.

Overall these alternative materials have a number of advantages, including:

- Lower cost in practice, in spite of higher raw material costs.
- Reduced evaporation losses due to their lower volatility.
- Higher efficiency due to improved solvent power.
- Some are more easily recyclable.
- Lower waste disposal costs for spent cleaning agents as they are less hazardous.
- Solvents of lower volatility are in general less toxic.
- Water-based systems can often be processed and discharged by consent to main sewers.

Water-based systems are usually neutral or alkaline and their use as substitutes for organic solvents is on the increase. They can be very effective as cleaning agents in steam systems and in high and low pressure spray methods. They are particularly good at removing water-based machining coolants, as well as chlorides and insoluble bulk contaminants such as dirt, grit and grease.

The water used in cleaning, especially that from rinsing, can often be recycled and re-used. If it is to be discharged to the sewer it may need to be treated in advance, depending on the type of contamination and additives present.

To obtain the best coating performance the water used at the rinsing stage is preferably de-ionised and filtered to prevent dissolved and suspended solids precipitating on to the surface to be coated.

Neutral aqueous solutions are excellent for use in spray and ultrasonic methods. They can also be used in hot water and steam cleaning equipment, giving good removal of solid matter, light oils, and chlorides and other salts.

Neutral solutions containing surfactants, corrosion inhibitors and other additives are best when a high degree of cleaning power is not required. They are not particularly suitable for immersion processes unless accompanied by vigorous agitation.

Vapour degreasers designed for organic solvents that do not meet current environmental standards often need only minor modification to convert to neutral aqueous solutions. Potential corrosion problems are overcome by modifying the process or by adding corrosion inhibitors. A drying system will also be needed. The hardness of the incoming water should be examined and treated if necessary.

Acidic aqueous solutions are effective for removing rust, scale and oxides. The acid and additive contents are tailored to the metal being cleaned and the type of contaminants present. The solutions may contain mineral acids, for example hydrofluoric acid, sulphuric acid, phosphoric acid or nitric acid. Chromic acid, and organic acids such as acetic and oxalic acid, are also used. Solutions may also contain detergents, chelating agents, and small amounts of water-miscible organic solvent to improve their effectiveness.

Alkaline aqueous solutions are used in all types of liquid processes for cleaning metal, glass and plastic components. They are good for removing grease, cutting oils, shop dirt, and particularly fingerprints.

The process gives a high level of cleanliness if the solutions are well filtered and are properly rinsed off afterwards. A wide variety of additives are used to improve cleaning, including sequestering agents, emulsifiers, and surfactants. It is often valuable also to include corrosion inhibitors with some metals, especially on aluminium and some non-ferrous alloys.

CORROSION

Corrosion shows itself in many ways, and there are indeed a number of different types of corrosion. Fortunately we find that if we are able to control one type of corrosion we often discover that other types are also prevented, or at least slowed down.

The exception is corrosion caused when two different metals are present. In areas of stress inside a metal substrate corrosion will actually be speeded up.

One of the many benefits of good pre-treatment is of course that it helps to reduce corrosion. We have seen that it can prevent surface defects and promote adhesion, but it also adds to the life of coatings by itself providing a barrier to corrosion. This is the case on both ferrous and non-ferrous substrates.

For corrosion to take place the main ingredients are moisture and air. Everyone is of course familiar with the rusting of iron and steel, which normally takes place quite slowly. If the atmosphere is polluted, for example with sulphur dioxide, the oxidation is much quicker.

The corrosion process is almost always electrochemical in nature and is produced by the formation of a cell, similar to that in an electric battery, made up of water, oxygen and dissolved salts. Its action is fast enough on the surface of ultra-clean steel to see it actually happening.

The use of a powder coating can provide an excellent barrier to this process, as long as the substrate is corrosion-free when it is applied.

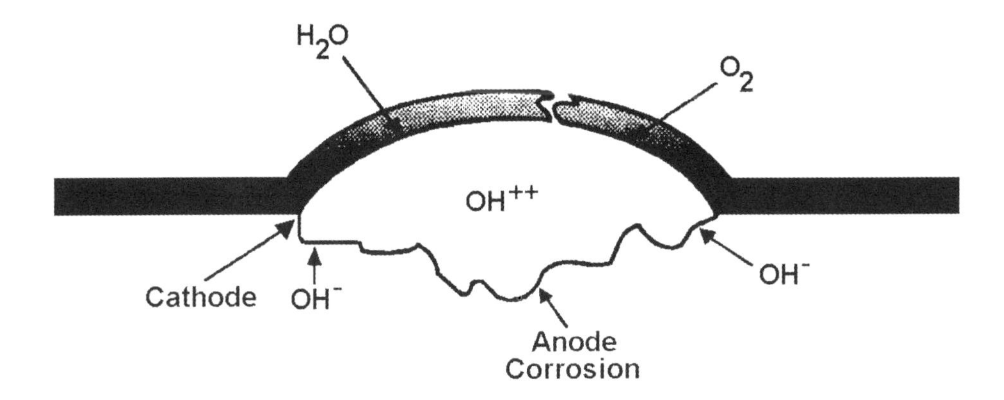

A corrosion cell

When a paint film is damaged and the underlying substrate exposed, and a combination of water, air and salt are present, then corrosion will inevitably take place.

Conversion coatings

Metal surfaces are treated with conversion coatings to increase the corrosion resistance of the surface. This category includes phosphating and chromating as well as alternative treatments now being introduced for environmental reasons.

Phosphate pre-treatment prevents corrosion occurring, or at least it slows down the process. This is due to the insulating effect of the phosphate layer, and not surprisingly the corrosion resistance improves with the weight of phosphate deposited. Unfortunately as it increases the adhesion characteristics of the coating are correspondingly reduced, and it is necessary to strike an acceptable balance between the two.

A similar principle applies to chromate conversion coatings applied to non-ferrous metals such as aluminium.

Ferrous materials are normally cleaned and phosphated followed by washing, passivating and drying, and finally powder coated. This process involves several stages, carried out either by immersion in a series of tanks or by spraying in a tunnel, for example:

- Degrease using an alkaline water-based degreasing solution.
- Rinse with water to remove any chemicals remaining.
- Apply the phosphate conversion coating.
- Rinse with water.
- Rinse finally to passivate or to remove salts, using de-ionised water.

The temperature and strength of the chemicals is important in ensuring that the correct weight of phosphate coating is applied.

Non-ferrous materials are chromated in a similar fashion. However, before this takes place an alkaline degreasing and etching step may be required, followed by 'de-smutting' with nitric acid solution. Acid cleaners can be used to avoid these additional processes.

In all cases routine control of the process is essential to ensure that proper quality is acieved.

Plastics and glass may also be treated chemically in multi-stage processes to provide a clean, grease-free surface, with a 'key' such as an etch to assist adhesion.

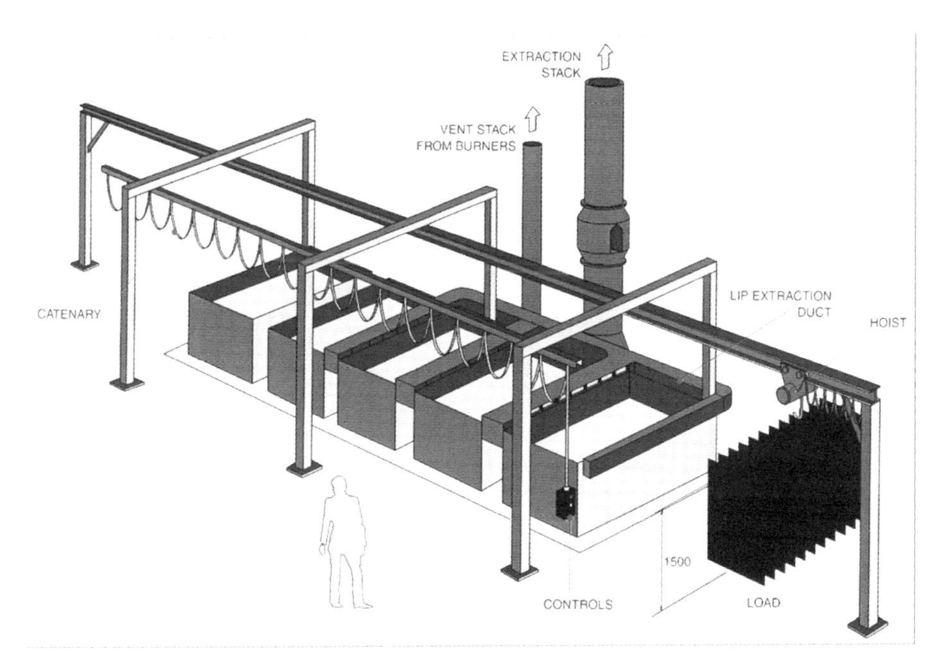

In-line phosphating

There are many variations on the systems described but they all have the same objective of achieving high quality in the final coating. There are a few common problems, including those mentioned below:

- Energy losses during heating and cooling of the tanks, particularly in spraying systems.
- The large amount of water used for rinsing.
- Treatment agents carried over into the following stages, including the rinsing.
- The strict control required in the process to maintain quality.

Cost savings can be achieved in all of these areas by attention to the design, operation and management of the treatment system in order to:

- Maximise productivity.

- Reduce waste.
- Control the balance between the strength of the chemicals, spraying or immersion time and temperature.

Phosphating

The object of applying a metal phosphate layer to the surface of a substrate is two-fold:

- To help the adhesion of the coating.
- To prevent corrosion, particularly if the coating is later damaged.

The phosphate layer promotes adhesion due to the fact that the surface area is effectively increased due to its fine crystalline structure. Coating powder will also flow into the micro-fissures of the phosphate layer, as its surface is readily wetted by the molten powder.

Some polymer systems even combine chemically with the phosphate layer.

Phosphating is generally applied to ferrous substrates, but non-ferrous components such as aluminium and zinc in the form of hot-dipped galvanised and plated surfaces and die-castings can also benefit, though the details will depend on the metal involved. It is found difficult In practice to treat different types of metal in one operation.

It is unusual for the treatment process and materials to be designed by the user; more often than not the user selects and purchases a proprietary system. The chemistry of the process is often made out by the suppliers to be very complicated, but it does in fact follow some simple principles.

The coating presented for powder coating is usually composed of iron or zinc phosphates, or occasionally manganese phosphate. Interaction between the substrate and the metal phosphate results in the formation of a fine crystalline layer of between 0.4 and 2.0 g/m^2, which is believed to be the optimum for adhesion and corrosion resistance. Above this coating weight the adhesion will become progressively poorer, though anti-corrosion properties continue to increase. The actual thickness of the phosphate layer can be up to 10 microns.

After the cleaning operations and a rinse to prevent contamination of the metal phosphate tank, the phosphate solutions will form secondary and tertiary salts along with free phosphoric acid. Process time, temperature and the equilibrium between the constituents of the process must be carefully balanced to produce the ideal coating weight and structure to achieve the results required.

$$3M(H_2PO_4)_2 \rightleftharpoons 3MHPO_4 + 3H_3PO_4$$

$$3MHPO_4 \rightleftharpoons M_3(PO_4)_2 + H_3PO_4$$

$$3M(H_2PO_4)_2 \rightleftharpoons M_3(PO_4)_2 + 4H_3PO_4$$

Three basic equations of phosphating action

Adding salts of nickel or copper, or a variety of oxidising agents, nitrites, nitrates and similar materials can speed up the action of the solutions.

The process can also be made quicker using spraying rather than immersion.

When non-ferrous materials are phosphated it is necessary to include fluorides to precipitate out unwanted dissolved salts. In the case of aluminium substrates it will be necessary to remove these precipitated salts by a 'de-sludging' process, as there can be a rapid build-up that can cause coating problems.

In every case there is a need to control the phosphate conversion coating process carefully so as to:

- Provide the ideal crystalline structure for adhesion of the coating.

- Reduce porosity and enhance corrosion resistance by ensuring fine and uniform crystallisation.

- Minimise process costs by optimising the relationship between time of processing, balance between the chemicals used, and temperature.

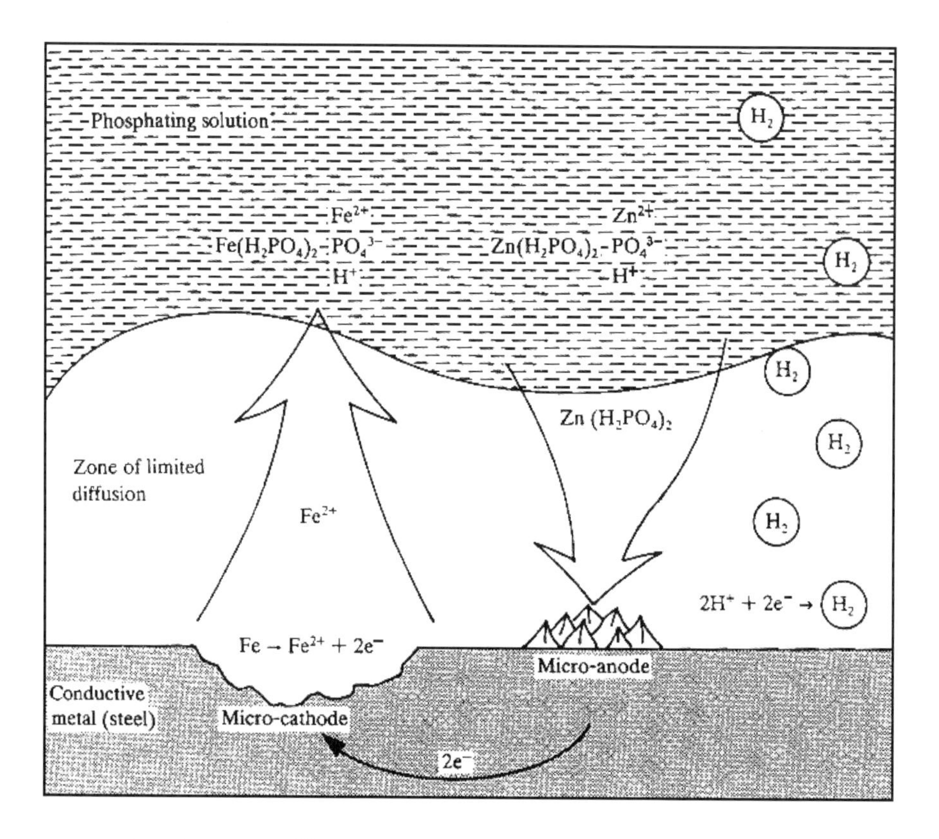

A schematic illustration of the phosphating process

Passivation of phosphate layers

It is common to *passivate* or seal the substrate after a phosphate conversion coating has been applied.

At the very least it is important to ensure that the final rinse of the phosphating process is carried out using de-ionised water so as to prevent deposition of salts suspended or dissolved in the water. The deposit of contamination such as this during the drying process can ruin all the good work put in earlier and can introduce defects into the final powder coating film.

Often a passivating chromate rinse is incorporated before or as part of the final rinse. It is thought that this improves the corrosion resistance of the phosphate layer by sealing its surface structure.

The use of chromium compounds is however becoming less acceptable environmentally and alternative materials such as molybdates and organic sealing agents are being used in the final stages of the phosphate process.

Combined cleaning and conversion

Combining the two into one process can save space. Two procedures are possible.

Some systems use a batch-type, multistage spray cabinet, looking rather like a large dishwasher, to clean components using a neutral or alkaline aqueous solution, followed by phosphating and rinsing. The need for solvent-based degreasers is therefore eliminated.

Other systems combining degreasing and phosphating use methylene chloride, or preferably inert solvents with slower evaporation, to degrease the components while they are immersed in a dip tank in which the phosphate layer is applied. An inert liquid, sometimes water-based, sits on the surface of the solvent to prevent the escape of organic solvent.

Such systems are especially suitable for continuous conveyorised systems and for handling large batches of small to medium-sized parts.

Another simple approach combining cleaning and phosphating is to use the first tank for both cleaning and phosphating and a second tank to finish off the phosphating process. This reduces the consumption of materials by using the uncontaminated iron phosphate solution to replenish the cleaning bath solution. Only a final rinse is required to remove chemicals and to passivate.

The advantages of combined degreasing and phosphating are clear:

- Less space is needed.
- Handling time is reduced.
- Capital cost is lower.
- Fewer chemicals are used.

The quality of the conversion coating needs to be monitored to make sure that sufficient cleaning has taken place early enough in the process to ensure that phosphating is adequate. The process is most often used for slightly soiled components.

Combined static degreasing and phosphating system

Conveyorised processing

This is especially suited to continuous processing or to handling large batches of small and medium sized parts.
Usually the same conveyor is used, non-stop, to convey parts through the Cool-Phos system and the painting or other coating stage.

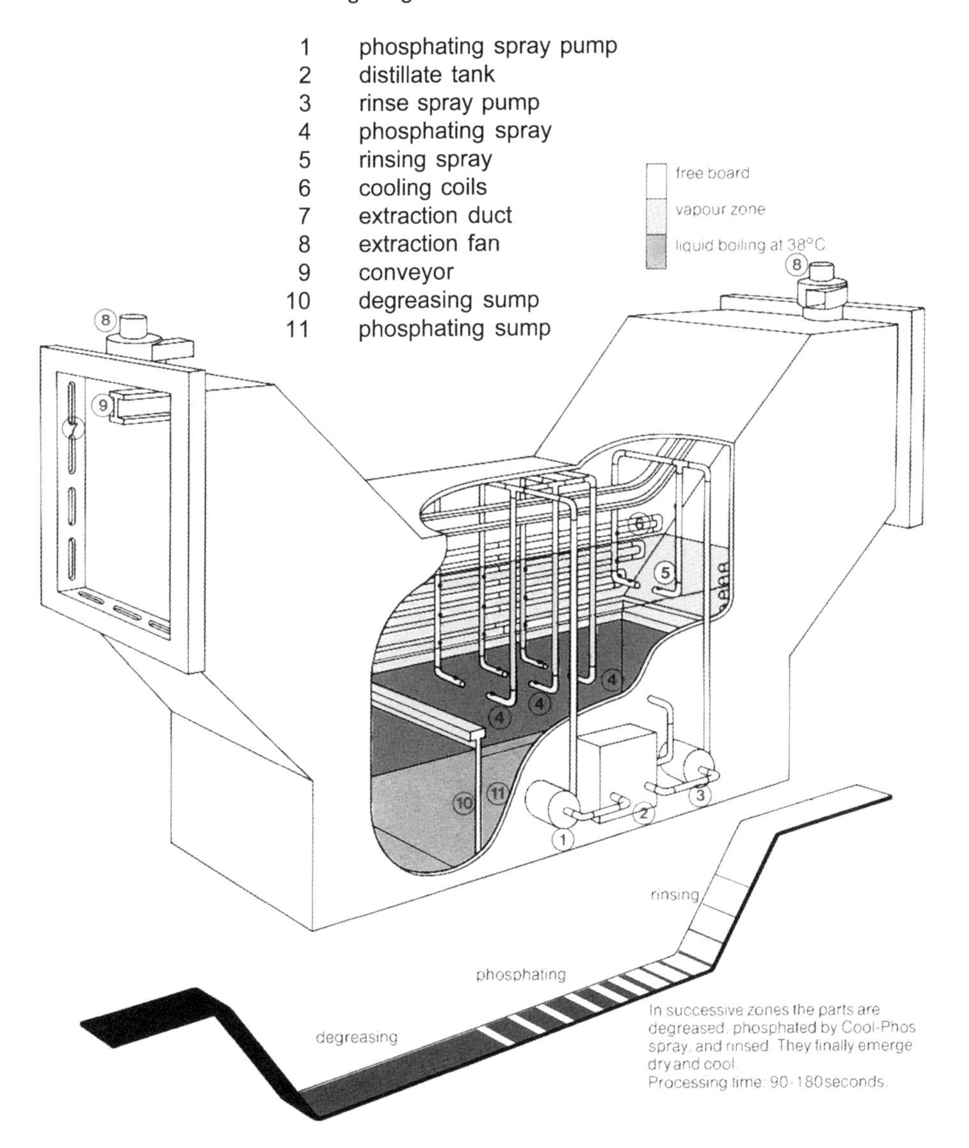

Processing by conveyor

Chromate conversion coatings

There are two types of chromate conversion coating used on non-ferrous substrates, based on either chromium phosphate or chromium chromate.

They are easily distinguished, as chromium phosphate imparts a green hue to the surface whereas chromium chromate is yellow-brown in colour.

The coatings are applied by immersion or spraying, as with phosphating.

Proprietary materials are always used and in general will contain phosphoric acid and chromate salts, with a fluoride accelerator in the case of chromium phosphate. In this instance the fluoride is present as hydrofluoric acid and close control is required to prevent etching of the component's surface.

The deposited layer is non-crystalline and should ideally be between 0.5 and 5.0 mg/m^2 in film weight for good powder coating. Heavier deposits in excess of 10 mg/m^2 will provide excellent corrosion resistance but the adhesion properties of the powder coating may suffer.

It is possible to apply this type of conversion coating either by immersion or by spraying, though for safety reasons the immersion process is usually preferred. It is quite universal in application and more dilute versions can be used as a post-rinse for ferrous materials.

The use of chromium chromate treatments is broadly specific to aluminium substrates, though they are also occasionally used on galvanised surfaces.

Accelerators are used in the process and the traditional ferricyanide additives have given way to molybdenum and selenium salts. The structure of the deposited layer is in fact a mixture of chromium chromate and aluminium oxide.

$$3Al + 6H_2CrO_4 + 6HF \rightarrow AlF_3 + Al_2O_3 + Cr_2(CrO_4)_3 + CrF_3 + 9H_2O$$

$$AlF_3 + NaF + 2KF \rightarrow K_2NaAlF_6$$

A coating weight between 0.5 and 2.0 g/m^2 is ideal for powder coating applications.

These chemicals are hazardous both to the user and to the environment. Particular care needs to be taken in use as they can cause dermatitis and operators can become sensitised by the chromium salts.

Effluent treatment

The use of chemical cleaning followed by a conversion process is universally accepted as the ideal way for the treatment of metallic components before powder coating.

However, waste is generated both from the contamination removed and from spent chemicals from the chemical processes. Chromium salts must be recovered from the effluent before it is discharged to the main sewer.

It is essential to think carefully about the disposal of these waste materials. As they are all either suspended or dissolved in the waste water, consultation must take place with the local water authority as to what levels are acceptable for discharge to the sewer. It will then be necessary to put in some kind of treatment plant to meet the standard agreed.

In some cases wastes will have to be drummed up and stored for removal by a specialised waste disposal contractor.

The wastes that may occur can be summarised as follows:

- *Organic solvents* that have been used to dissolve or emulsify contamination on the substrate can in some cases be recovered in-house. Otherwise they must be removed from the site and disposed of by a specialist contractor.

- *Acidic solutions* will have to be neutralised and the salts precipitated out for disposal. These salts often contain chromium as well as less toxic elements such as calcium, zinc and aluminium. Once again an authorised contractor must undertake their disposal. It will often be necessary to carry out an analysis to verify the composition and thus the degree of hazard.

- *Alkaline solutions* can be treated in a similar way to acidic waste by neutralisation and precipitation. Unfortunately many spent cleaning solutions of this type contain high quantities of surfactants and other agents that hinder the precipitation and filtration procedure.

- Heavy metals, phosphates and chromates require more thorough treatment, as their disposal is restricted.

In all cases the effluent plant and the responsible use and disposal of water should be considered an integral part of the chemical pre-treatment process, not simply an optional extra.

OTHER PRE-TREATMENTS

Other treatments are often carried out before powder coating to improve the quality of the final coating. Most of these operations are performed on a case by case basis, either because an effective general treatment is not available, the components are too large, or the quantity to be coated does not warrant the installation of a custom-built pre-treatment plant.

Phosphoric acid washing

It is useful to apply a wash based on phosphoric acid to hot dipped galvanised steel before powder coating. The coating is usually thinly applied by brush and allowed to dry. A 'mordant wash' of this type can help to provide a uniform surface for powder coating.

The application should be light and the material dilute.

Pigmented etch primers

An ideal surface for powder coating can be provided by the application of a conventional two-pack self-etch chromate pigment primer. Single-pack etch systems can sometimes be adequate but the two-pack type is safer and more resilient. It is a solvent-based material and is usually applied by spraying to obtain a light and even coating.

The film must be thoroughly dry before the coating powder is applied.

Etch primers provide an excellent surface for powder coating and improve adhesion and corrosion resistance. They are ideal for aluminium and other non-ferrous materials such as galvanised surfaces and zinc die-castings.

Powder coaters have also occasionally been known to use them on ferrous substrates.

Primer coating powders

We have already seen that powder coating, being a single-coat application, can give poor corrosion resistance on ferrous substrates if damage to the coating occurs. Certain critical applications call for the highest corrosion resistance, and it is interesting that some of the principles used in the application of paint have been adapted to powder coating to meet this requirement.

The application of a zinc-rich coating powder as a primer, which is then melted but not fully cured, can offer an excellent base for the final decorative top-coat.
Such systems at a total thickness of 100 -120 microns have been found to pass the 2000 hours neutral salt spray test, ASTM B117.

Electrophoretic primers

Many powder coating systems, particularly those for the automotive industry, apply an electrophoretic primer before coating.

The principle is that after pre-treatment the metallic component is dipped into a bath of paint and an electric current applied, the component becomingone of the electrodes. This causes the coating to be deposited on to the substrate in a very efficient and uniform manner.

The primer coating has to be fully cured before the powder coating is applied and it then produces excellent corrosion resistance.

Such systems are sufficiently corrosion resistant to pass with ease the 1000 hours neutral salt spray test, ASTM B117.

CHAPTER IV

POWDER COATING APPLICATION

Chapter IV

POWDER COATING APPLICATION

THE CHOICE

There are a variety of ways of applying coating powders to different substrates. In each case an assessment has to be made of the quality, coating performance and cost requirements. The choice of the procedure to be used will depend on the answers to the following questions:

- Why is the coating being applied?
- What is the substrate?
- How many components are involved?
- What performance has been specified?
- What coating thickness is needed?
- What colours are wanted?
- What are the production requirements?
- What processing costs are acceptable?
- What is the cost of the equipment?

We will look at these one at a time. In practice the choice is not simple as the important factors are always interrelated. Some may be mutually conflicting, for example the constant demand for high quality at low cost, but usually one or two factors will stand out as the dominant issues.

Why coat?

There are broadly three reasons for coating, though more than one of these may apply at one time:

- Appearance – a coating is applied for decorative effect.
- Protection – it is needed to give a protective barrier against impact, wear or corrosion.
- Special properties – for example, anti-friction, electrical resistance, anti-graffiti, etc.

The list of special properties list gets longer as time goes by, and as coating powder technology advances to offer more and more benefits to the end-user.

What substrate?

The development of new materials and curing methods is rapidly extending the range of substrates that can be powder coated:

- Ferrous metals.
- Non-ferrous metals and their alloys, particularly aluminium, but now including such unusual metals as magnesium and titanium.
- Plastics – can now be coated using low-bake and UV-cured materials. This will be a big growth market in the future.
- Wood and MDF composites – can be coated with low-bake powders. Mass-produced furniture manufacturers are seeing this as a cost effective solution to their environmental problems.
- Glass – an ideal base for powder coating, both for protection and colouration. It is possible to reduce the weight of glass needed if it is powder coated, with the added bonus of improved impact resistance. In all cases the substrate must be able to withstand a certain amount of heat. Low-bake powders are now available that can be cured as low as 120°C.

Higher temperatures are involved in some processes, particularly during pre-heating for fluidised bed coating. For example, wire work is often heated to around 400°C so that it is able to pick up the powder and still retain enough heat for it to flow out.

How many items?

The quantity of components to be coated will decide whether a manual or automatic application method would be best. Other factors to be taken into account are:

- How big they are
- Their complexity
- Their configuration

What is the specification?

The make-up of the coating formulation and how it will be applied depend very much on the performance required by the customer.

User requirements fall into one of the following broad categories:

- External use, perhaps for architectural applications.
- Internal uses, with additional requirements such as appearance, surface feel and abrasion resistance.
- Chemical resistance. This may vary between normal use in a domestic situation, and the ability to withstand sterilisation in a hospital or strong acids in a chemical plant.
- Abrasion and impact resistance. This almost always applies to some extent.

How thick?

Some application methods can only provide thin films, while others will only give the opposite. The coating thickness applied can range between 50 microns or less, up to 500 microns or even more.

What colours?

If just one colour is wanted, the choice of coating method is quite simple.

Life becomes more complicated when several colours are required and when some of the polymers to be used are incompatible with others, as this can create contamination problems in processing if the wrong choice is made. The number of coatings and colours to be applied very much affects the application method selected, particularly when rapid colour changes are likely to be involved.

Production factors?

In addition to the problem of colour changes, a big issue is the coating speed needed to meet the production target. A calculation has to be made for each type of application being considered in order to be confident that the production rate over a given time can be achieved, maybe with some scope for improvement in the future.

Processing costs?

To calculate process costs the *transfer efficiency* of the powder in the application process needs to be established. Some processes give almost 100% powder utilisation whereas others may only give 25%. This is particularly true for spray processes if powder needs to be sprayed to waste due to the rapid colour changes required.

Careful thought has to be given to the cost of reclaiming waste powder. Recycling of powder is simple in principle but not without its problems, as contamination, particle size balance and eliminating static charge are all issues. It is clearly much better to concentrate on transfer efficiency, as the higher we can make it the less powder there will be to re-process and the less need for virgin material to be added.

What will the equipment cost?

Installing and running a powder coating plant is not cheap. In its simplest form it can be just a simple fluidised bed plus a small oven, but to be a player in the big league a fully automated electrostatic powder coating plant with conveyorised pre-treatment and oven is going to call for serious investment. The difference between the two in labour and energy costs can also make a big difference.

The final choice

When all the alternatives have been looked at, the choice will generally fall between the two primary application methods:

- Fluidised bed coating
- Electrostatic spraying

There are other processes that might be used to meet particular demands, but these usually boil down to a combination of one of these primary methods with other special techniques. It also possible to link the two techniques themselves, for instance by using electrostatically charged powder in a fluidised bed.

We will be looking at fluidised bed and electrostatic spray coating in detail.

Compared with wet paint application, powder coating has some obvious advantages:

- *No solvent*. The environmental benefit of this is counterbalanced to some extent by the need to control dust and to protect operatives.
- *Quicker processing*. With powder coating there is no waiting for solvent to flash off and in most cased a primer coat is not necessary; in both respects this shortens the process.
- *Lower rejection rate*. Powder coating is a straightforward process and there is less to go wrong.
- *Versatility*. A wide range of finishes is available to meet most decorative, corrosion resistance and other performance demands.
- *Heavy coating weight*. Enhanced protection and decorative impact can be offered without difficulty.
- *Ready to use*. Powder is used straight from the box, unlike paint which perhaps has to be adjusted in viscosity for spraying. It is also less messy.
- *Cheaper packaging*. Because powder coating gives thick tough films, components are less prone to damage during transportation and can be packaged more cheaply.
- *Reduced energy needs*. There are no solvents to be vented away from ovens and spraying areas, meaning fewer air changes and less energy usage.
- *One-coat process*. Up to 100 microns can be applied in one coat.

Powder coating also has some disadvantages:

- *Inflexibility*. Once manufactured coating powders cannot be altered or modified.
- *Film-weight*. High film thicknesses are in practice not always wanted,

and powder coatings can seldom be applied below 30-35 microns and still give a smooth opaque finish.

- *Corrosion Resistance*. Powder coating is normally a one-coat system and unless the preparation is really good, possibly including a primer, a multi-coat paint system may in fact be better.

FLUIDISED BED COATING

This is the simplest method of powder coating, dating back to the early 1950s when it was used to apply powders based on epoxies, cellulose acetate butyrate (CAB) and polyethylene. Nowadays more sophisticated materials are applied by this method, for example polyamides such as Nylon 11 and 12, PVC, and hybrid powders containing mixtures of polymers to meet specialised end-user demands.

Fluidised bed coating is generally used for applying thermoplastic materials as these do not require to be cured afterwards. Processing costs are competitive since there is only one heating process before coating. If a single pre-heat cycle is for some reason not possible another method of application will be chosen. This process can give high film thicknesses, greater than are easily obtained by other methods of application.

The coating powders normally have a particle size between 40 and 350 microns.

Fluidised bed equipment

The fluidised bed unit is made up of an open topped straight-sided tank fitted with a porous tile at the bottom. Air is blown upwards through the tile from a plenum chamber, as shown in the diagram, and causes the coating powder to dance around in a 'fluidised' state. The tank has to be of sufficient size to allow the components to be fully immersed and easily moved around within the fluidised powder.

The porous tile or membrane may be very simple, perhaps merely a number of layers of canvas, but nowadays it is more likely to consist of a panel of sintered plastic or porous ceramic. In large tanks containing several tons of coating powder the efficient design of the porous membrane is crucial for uniform fluidisation throughout the whole mass of powder.

When powder is added to the tank an allowance has to be made for the increase in volume that takes place once fluidisation begins. The volume of the coating powder may be found to increase by half as much again.

The air supply passing through the tile should be controlled to avoid too much turbulence, so that the fluidising tank looks to be gently simmering rather than boiling. This is best managed using a large volume of air at low pressure rather than the other way round, and it is obviously better to supply the air from a high-volume low-pressure blower rather than using compressed air. In any case the air blown in must be clean and free from contamination, particularly oil and water.

The fact that fluidised powder exists in suspension makes it possible to dip components of virtually any shape into it. The circulating air allows the powder to flow like a liquid around complex shapes.

It is advisable to earth the tank as some coating powders can quickly acquire a static charge. In addition a local dust extraction system may be needed, since some coating powders contain a high proportion of fine particles that have to be vented away from the operators. In any case it is always advisable for operators to wear masks.

It will be clear from the illustration that the tank is simple in shape, avoiding areas that might trap powder or hinder it from fluidising effectively. If fluidisation problems occur the quality of the coating will quickly suffer.

The principle of the fluidised bed process is to heat the component so that the powder will melt on to its surface. When the component is removed from the bed it should still be hot enough to enable the powder to flow to a smooth finish, but not so hot that it begins to degrade. At this point, in the case of thermoplastic powders, the process is finished. If thermosetting powders are in use the coating will then need to be cured.

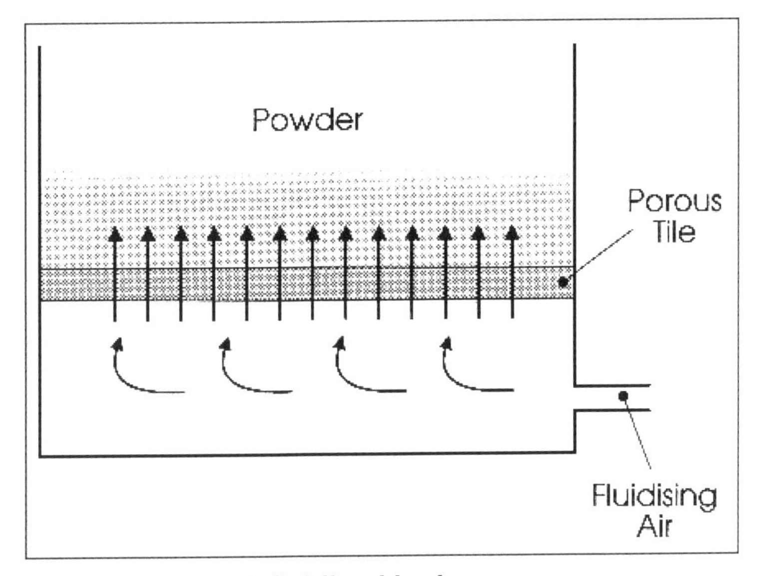

A fluidised bed

If particles of dry powder are still present and a smooth finish has not been formed, it is possible to reheat the components in the oven until they melt and flow out properly.

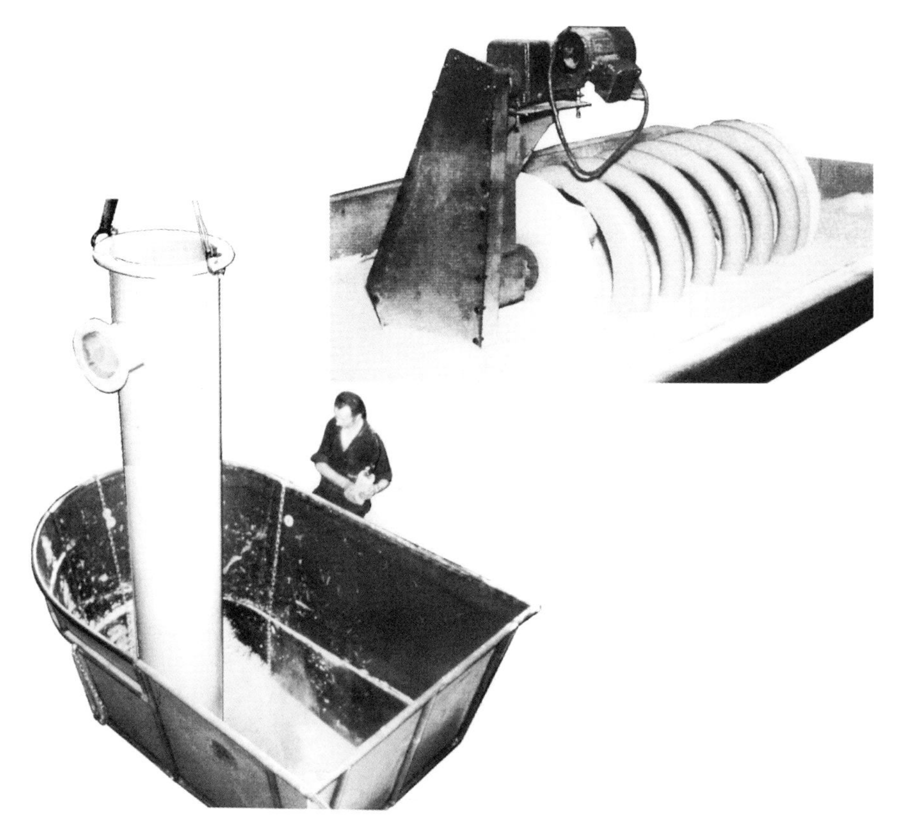

Dipping a component

Problem areas with fluidised bed coating are as follows:

- *Pre-heat temperature*. An optimum temperature has to be found at which the coating material is able to melt and flow without discoloration or degradation.

- *Differing thicknesses of metal in the component*. This can of course make it difficult to establish the ideal temperature.

- *Trapping of powder*. Some components may trap powder, and this can be difficult to remove. The dipping procedure, the jig holding the component and the way the component is actually immersed and withdrawn may need to be modified.

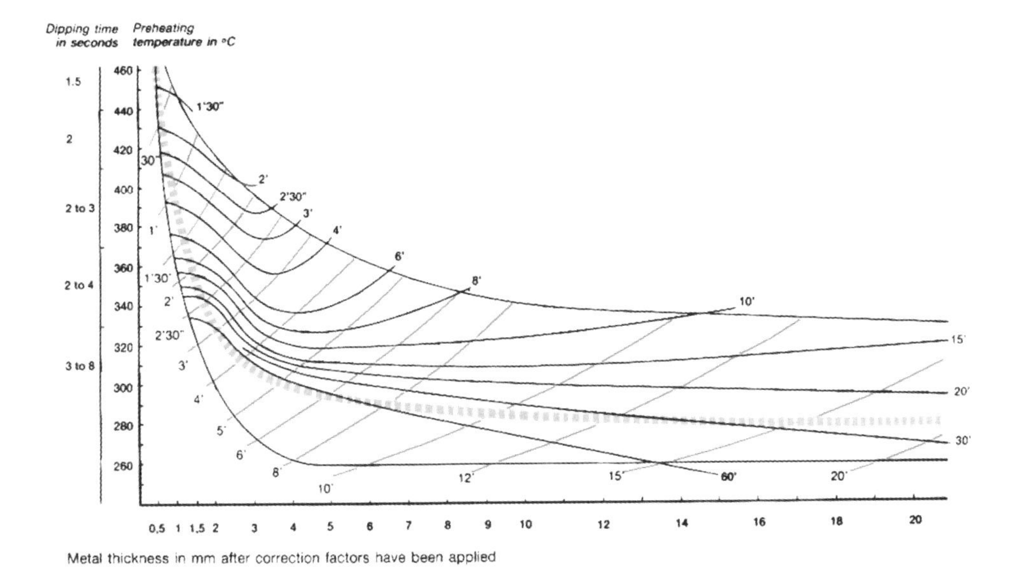

Correction factors -
Corrections must be applied according to the shape of the parts and the nature of the metal.

Shape of the part	Factor to be applied
tubes (external coating only) use the wall thickness	x 1.5
hollow objects (extinguishers etc.) use wall thickness	x 1.2
square profiles, wires or round iron rods use the flat side or on the diameter	x 0.4
flat or embossed sheets, flat shapes, T or U profiles, etc. (coating on both faces) **tubes** (internal and external coating)	x 1.0

Pre-heating temperatures for Rilsan™ Nylon 11

- *Masking*. It is often necessary for some areas to remain uncoated, and various methods of masking can be used. The jig may have pads and panels to mask the areas concerned. Alternatively, high-temperature masking tapes are used, or non-stick paints sprayed on and simply removed afterwards with a knife.

An illustration of pre-heating temperatures is given in the graph above. These are conditions required for Rilsan™ Nylon 11, a polyamide powder melting at 184°C.

A simple experiment can be carried out using different thicknesses of metal on a series of jigs. Each jig is pre-heated for different times to different temperatures and an infrared thermometer used to establish the actual temperature of the metal in each case. The jigs are then dipped into the powder for specified times and the results noted and analysed. This can be used to produce a ready-reckoner to work out the time and temperature required in the oven for a dip cycle based on the following variables:

- The thickness of the metal.
- The different metals in the component.
- The melting point of the powder.
- The time in the fluidised bed.

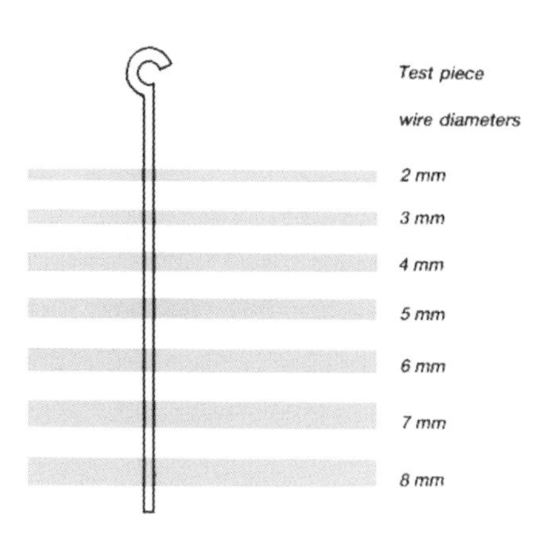

Table to be used in conjunction with the coating test piece

Wire diameter (mm)	400	380	360	340	320	300	280
8	VCY	CY	Y	SY	AS	AS	AS
7	CY	Y	SY	AS	AS	AS	AS
6	Y	SY	AS	AS	AS	AS	AS
5	SY	AS	AS	AS	AS	AS	AS
4	AS	AS	AS	AS	AS	AS	AS
3	AS	AS	AS	AS	AS	*	IF
2	AS	IF	IF	IF	IF	IF	IF

VCY very considerable yellowing
CY considerable yellowing
Y yellowing
SY slight yellowing
AS appearance satisfactory
IF incomplete fusion
***** a few unused particles

Example: Test results giving a ready-reckoner for wire coating

In general terms, a wire component may need to be heated for 10 to 15 minutes at over 300°C, so that it will be at about 250°C when it enters the nylon coating bed. If a dip time of just a few seconds is used, and with a quick tap on the jig to remove excess powder, a smooth coating will then result.

On the other hand a heavy valve might need to 'soak' in the pre-heating oven for 45 minutes at a temperature of 220°C before a thick coating of powder can be applied. In this case the dip cycle may be more leisurely, depending on the thickness actually required. All components will tend to lose heat rapidly in the fluidised bed due to the cooling effect of the air, so after dipping the valve would need to have excess powder removed either by rotating it or by using a low-pressure air gun. Sufficient heat should remain within the valve to obtain a smooth finish.

Water quenching

It is often useful to cool components rapidly after coating so that they can be processed more quickly. This is carried out by *quenching* them in a tank of cold water, a small amount of water softener being added to avoid 'tide-marks'.

One drawback is that this sometimes changes the appearance of the coating by reducing the gloss level. Another is that if air has for any reason been trapped under the coating, rapid cooling can suck water in and cause the film to peel off.

Post-heating and curing

Should additional heating be needed to help flow-out or to cure the coating, the temperature of the oven must be carefully regulated.

Thermosetting materials require a prescribed minimum time and temperature to cure, and overheating will cause degradation and discoloration. It may also make the coating sag and run, making it necessary to strip and recoat it.

FLOCK SPRAYING

If components require an extra-thick coating they can be reheated and recoated, a process known as *flock spraying*. This technique was originally used to spray fibres ('flock') on to a heated sticky substrate to give it a soft feel, but the term is now used to include spraying a further layer of powder on to an already coated article. It is used for large fabrications such as vessels, pipe-work and valves in the chemical industry.

The 'first coat' can be applied by the fluidised bed method. The subsequent coats, applied after each reheating, can be more selectively applied by a spray application using a flock spray gun. It is possible to apply coating thicknesses in excess of 500 microns.

CONVEYORISED FLUIDISED BED COATING

The fluidised bed method can be automated by *conveyorisation*.

How the components are handled during coating is very important, as some movement has to occur during the dipping cycle, as well as agitation after removal from the bed to remove excess powder. The jigging system must however be rigid enough to accommodate the flotation effect of the fluidised powder.

An additional step is sometimes added to the manual or automated fluidised bed procedure, as some components benefit from having a primer sprayed on to improve the adhesion and corrosion resistance.

Wire products are often coated in this way using a pre-heat, dip and post- heat procedure. Nylon-coated wire trays for dishwashers are a good example.

Conveyorised fluidised bed coating

ELECTROSTATIC SPRAYING

This is an area of rapidly growing importance in the product finishing industry as environmental pressures have an increasing impact. New techniques have to be introduced to reduce solvent and other chemicals being released into the atmosphere. This method is the ideal answer.

Electrostatic powder coating is known as a BATNEEC ('best available technique not entailing excessive cost') solution for companies which are trying to improve environmental practices and replace solvent-based systems.

It involves spraying on powder in a stream of air from a special gun, an electrostatic charge being applied at the same time. The charged powder is attracted to the substrate of the component due to the electrostatic force.

The coated component is then heated in an oven in the usual way so that the powder melts and flows, and then cures if it is the thermosetting type.

Electrostatic spray gun unit and spray booth

Electrostatic spray coating has a number of benefits:

- *Coating thickness.* Thinner films than fluidised bed application are the norm.

- *Powder recovery.* Over-sprayed powder can be collected in the spray booth, conveyed to a recovery unit and sieved for re-use. This is in many ways an ideal situation, with the theoretical possibility of 100% material utilisation. This is not the case with sprayed liquid systems.

- *Simple operator training.* The electrostatic principle is simple and easy to practically apply. It is very forgiving in practice.

- *Reduced energy costs.* Because powder is used so efficiently in electrostatic spraying the air used in the process area can safely be recirculated after powder recovery, reducing heating costs.

The process unfortunately also has certain disadvantages:

- *Quality control*. The final surface finish and gloss can only be assessedonce the item has been stoved, since before the melt process the surface finish is rather dusty and rough and gives little idea how it will eventually turn out.

- *Colour changes*. Thorough cleaning is essential between colours as contamination is a serious problem. The colours do not actually mingle but show as coloured flecks in the film.

- *Incompatibility*. As in the case of colour changes, time has to be allowed for cleaning down between one type of coating powder and another if they are not compatible. Contamination will result in coating defects.

- *Wastage*. Although powder over-sprayed in the application process can be recovered, the cost of the recovery process sometimes outweighs the value of the powder recovered, especially when the decision has been made to spray to waste between colours due to time needed to clean down.

Despite these minor drawbacks, electrostatic powder coating looks likely to be a winner for the future, but the change from wet painting has to be thought through carefully before taking the plunge.

The theory of electrostatic powder coating

Powder is conveyed to the spray gun in a stream of air. During spraying a static charge is applied from a high voltage source so that the powder becomes attracted to the component and sticks to it by electrostatic force. The coated component is then conveyed to the oven and the coating melted and then cured where appropriate.

Gravity naturally affects the powder between the applicator and the component, and the effectiveness of the application process very much depends on how efficiently the powder is charged.

The diagram illustrates the forces acting on a powder particle during its movement from the spray nozzle to the component.

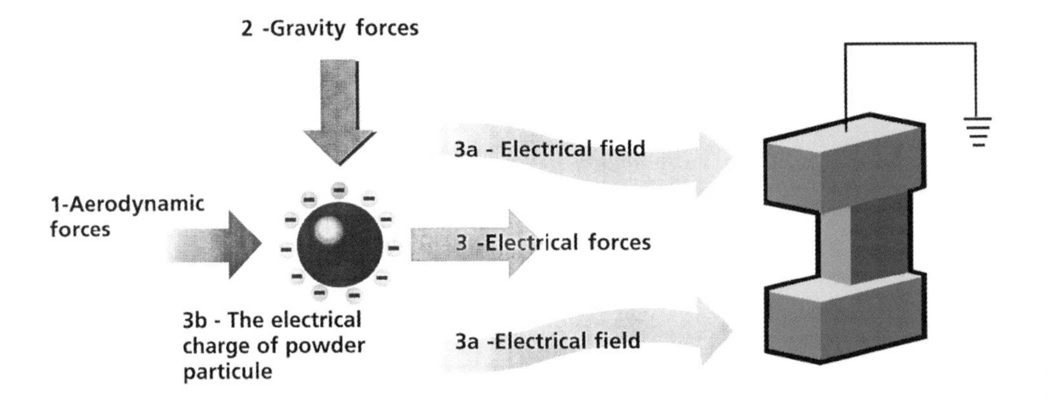

Forces acting on a coating powder particle

The charge is applied to the powder in one of two ways:

- From a high-voltage source emitted from a charging point. This normally imparts a negative charge to the powder and is called *corona charging*.

- By friction inside the spray gun, which gives the powder a positive charge. This is known as *tribocharging*.

Corona charging

Powder is conveyed in the air stream past an electrode having a negative potential up to 100 kV. A stream of ions in the air carries electrons to the powder, making it negatively charged and strongly attracted to the earthed component.

The polarity is occasionally changed to positive for some polymers like polyamides (nylons) which more readily accept a positive charge.

The corona charging point or electrode is normally supplied by a high-voltage source located in the barrel of the spray gun, but it can alternatively be delivered by cable from a source next to the application area. The electrode may be a needle at the end of the barrel, at which point the powder emerges as a cloud, or a group of small needles inside the barrel. These are respectively referred to as *internal* and *external charging*.

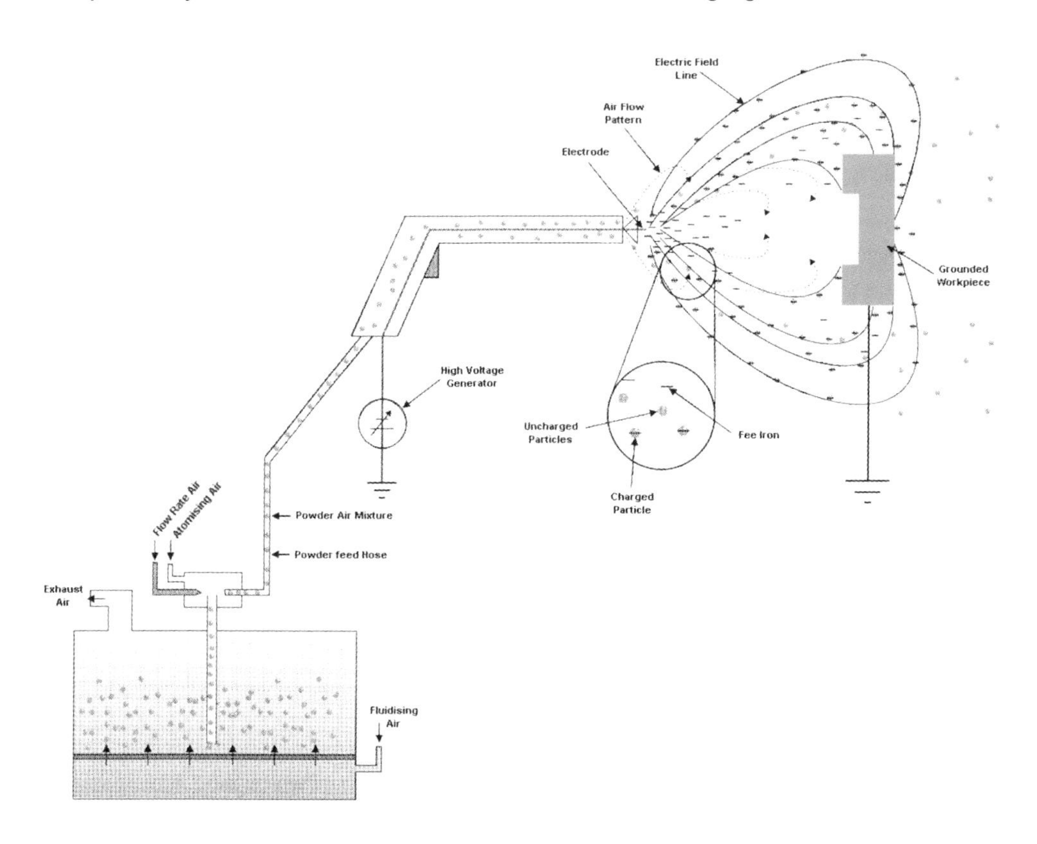

Corona charging

The corona process is the more commonly used and is fast and easy to operate. Film defects can arise, however, from variations in the level of charge on the powder.

A different problem arises if there is electrostatic interaction between the electrode and the earthed substrate. This has the effect of ionising (charging) the air stream with the same polarity as the powder and is known as *back ionisation*. If it is carried through to the coated film it causes mini-explosions within the coating as it waits to be melted, showing up as pock-marks in the final film.

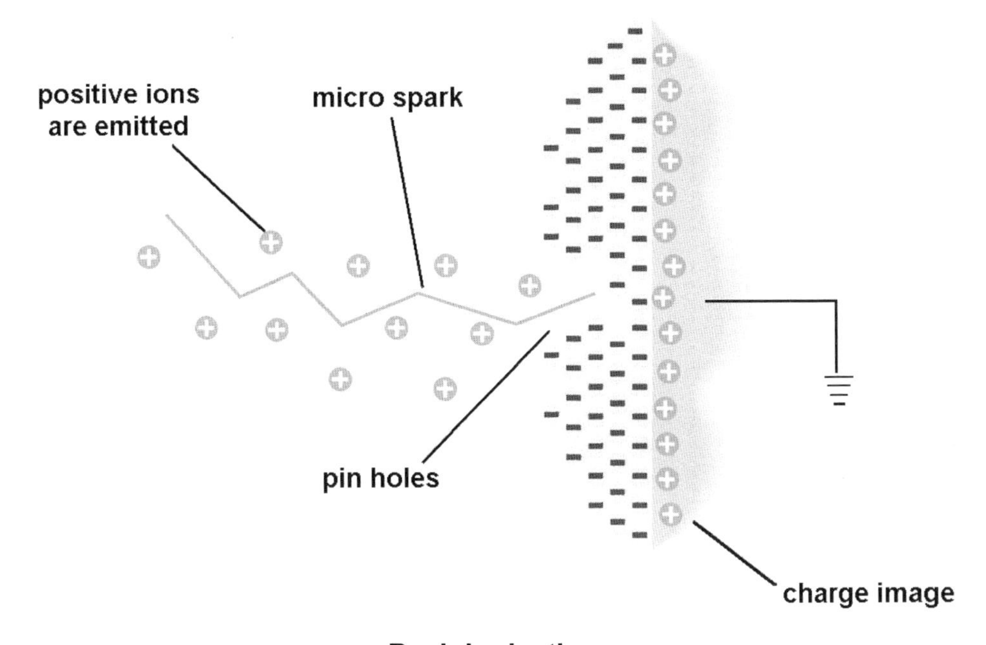

Back ionisation

Sometimes corona charging guns are fitted with 'free ion' collection devices to avoid these problems, for example by placing earthed needles outside the gun alongside the spray nozzle. They work by earthing the free ions in the area between the gun and the component, a process known as *ion stripping*.

These devices will often help the powder to penetrate into enclosed areas, as well as giving smoother and more even coatings, but they have the drawback of reducing the transfer efficiency of the coating powder.

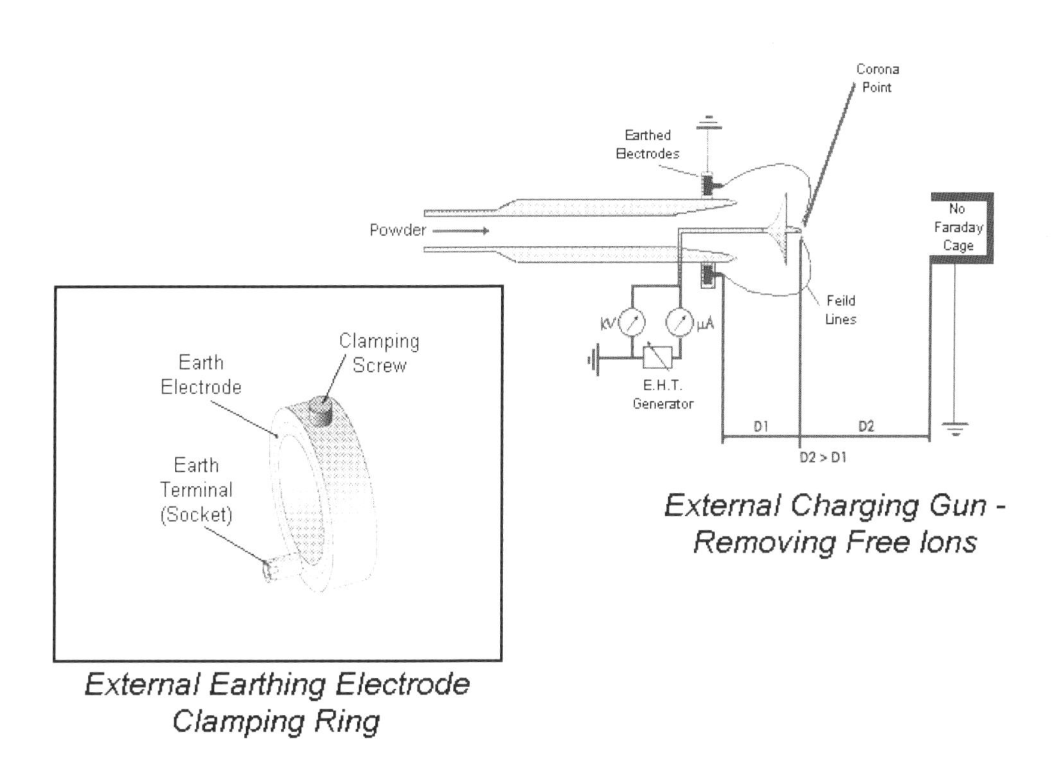

External Earthing Electrode Clamping Ring

External Charging Gun - Removing Free Ions

Effect created by an ion-stripping device

Tribocharging

This method of charging is quite different and depends on the well-known phenomenon that static electricity can often be generated when two different materials are rubbed against one another (*tribo* = friction). This is illustrated by the familiar children's experiment involving comb and paper.

In electrostatic powder application equipment the powder feed tube in the spray gun is made of PTFE, and friction between this and epoxy or polyester powders passing along the tube causes them to become positively charged. They are then attracted strongly to the earthed component.

It has to be recognised that some powders are more readily tribocharged than others and the powder manufacturer adds special ingredients to the coating powder to enhance the effect. In practice some of the powder particles may remain uncharged, but importantly there are no free ions in this process.

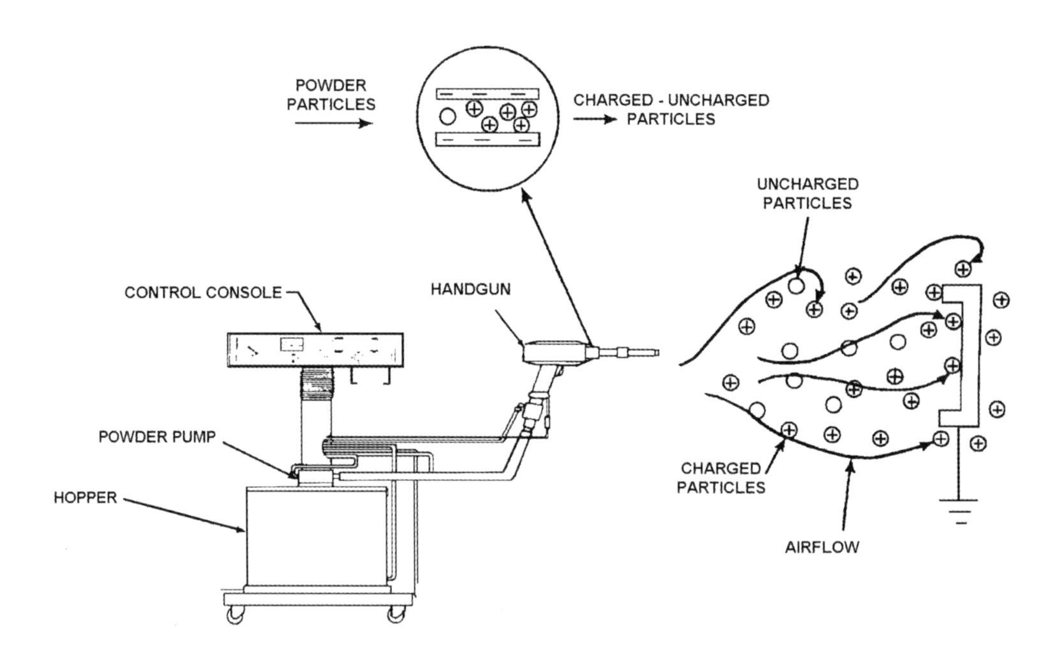

Tribo charging

Since there is therefore no danger in this process of unwanted ionisation, the charged powder particles are deposited evenly over the substrate and this avoids many of the film defects that occur with corona charging. Tribocharging is however generally slower than the corona method of application.

The *Faraday cage effect* is a term associated with electrostatic coating and relates to the difficulty of spraying into enclosed areas. Corona-charged powders with excess free ions in the powder cloud find it difficult to penetrate into enclosed corners. This is of course eased by the use of an ion-stripping device but the tribocharging technique avoids the problem altogether.

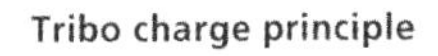

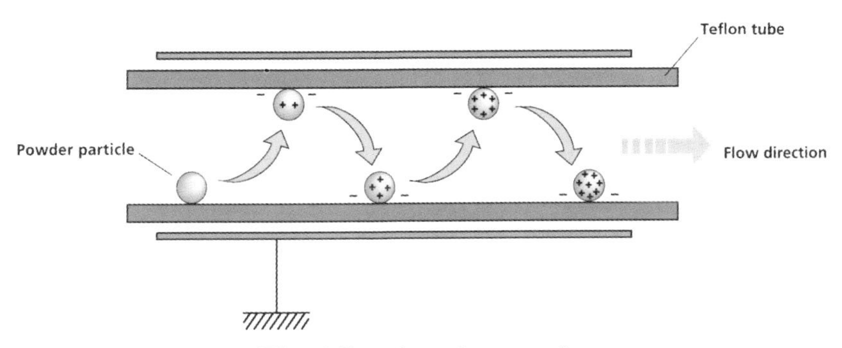

The tribocharging system

Manual electrostatic application

This is the most widely used powder coating application system in industry today.

Powder as supplied by the manufacturer is transferred to the powder feed system, enabling it to be delivered in a flow of air to the hand-held spray gun. Over-sprayed powder is collected in an arrangement that separates the air from the powder and allows it to be recovered for re-use.

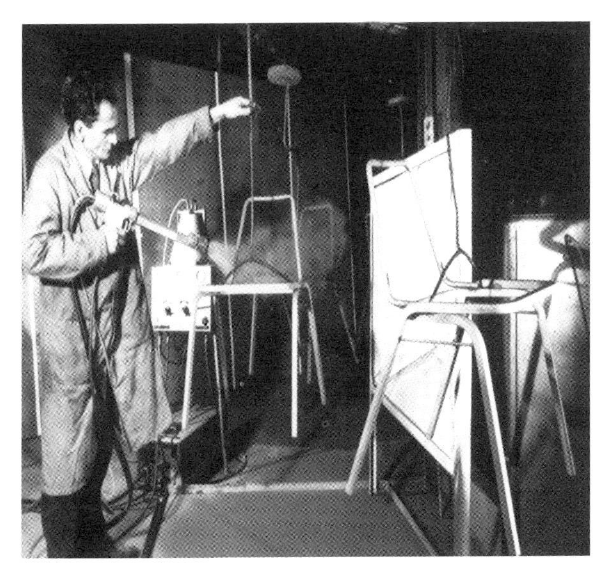

An early manual spray gun and booth

The essential components of the equipment are as follows:

- A method of feeding powder to the application device.
- A manual spray gun.
- A booth in which the component is sprayed and excess powder collected for recovery, designed to protect the operator from exposure to dust.
- An oven to melt the powder and to cure thermosetting types.

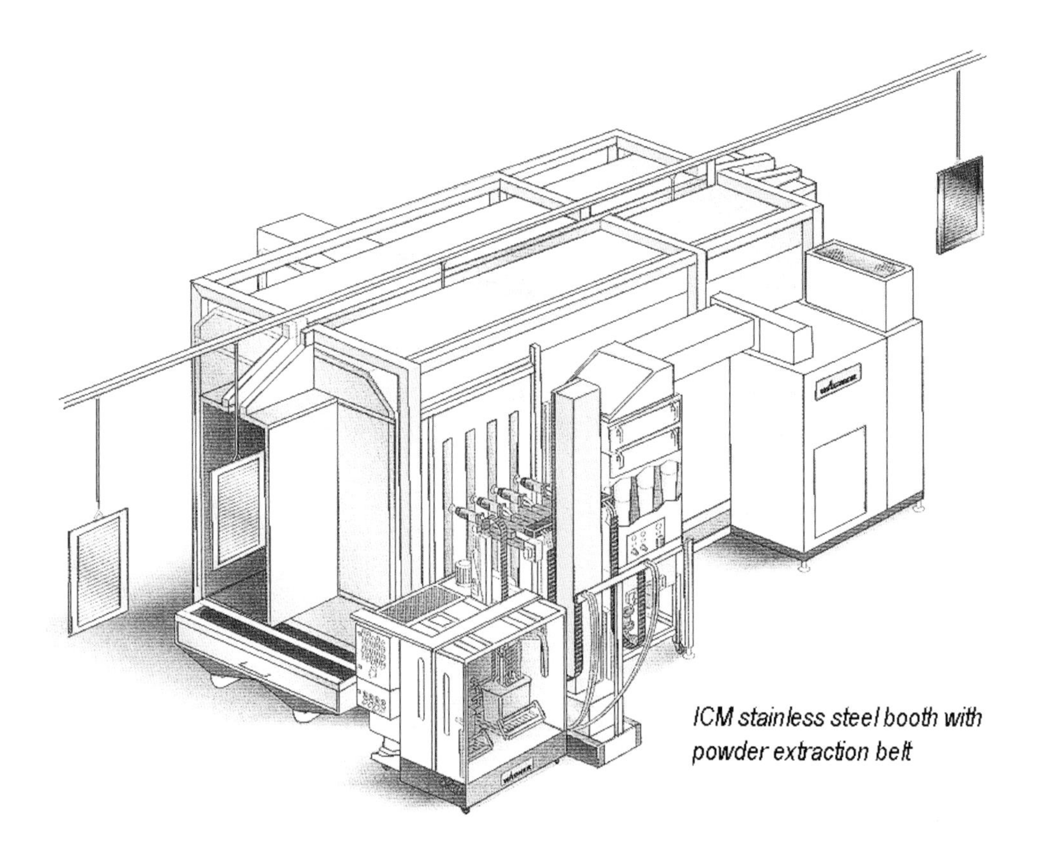

An automated electrostatic powder coating system

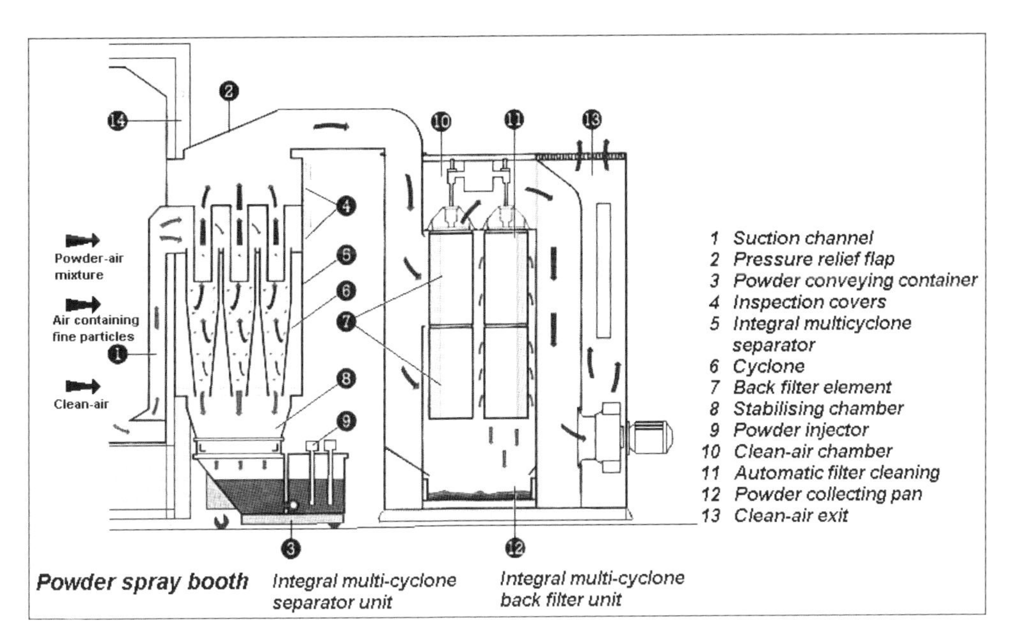

The operating principles of an electrostatic powder coating system

Powder feed systems

If uniform defect-free coatings are to be achieved, control of the flow of powder to the application gun is critical. There are three components to the powder delivery chain and each has an essential role to play if the system is to operate efficiently:

- The coating powder reservoir, comprising a fluidised bed, box or hopper.
- The powder pumping unit, provided with a venturi.
- The application device, normally a spray gun.

Fluidised bed hopper

This unit holds the powder, fluidised by agitating it vigorously while compressed air is blown through. The unit is similar to that used in fluidised bed coating but smaller. The powder hopper has to be the right capacity for the application it is supplying, which can mean several tons in a large-volume automatic plant.

An early integrated fluidised bed hopper

Early units used in powder coating resembled liquid paint pressure-feed tanks and operated batch-wise. Powder was held in a pressurised hopper fitted with a porous fluidising base and the unit vibrated. Powder was ejected from a variable-flow collection device in the lid of the hopper into the powder feed hose. This method gave an even and well controlled powder supply. The main drawback was that when topping up the process had to be stopped to allow the pressure to equalise. The units had limited powder capacity and were difficult to clean.

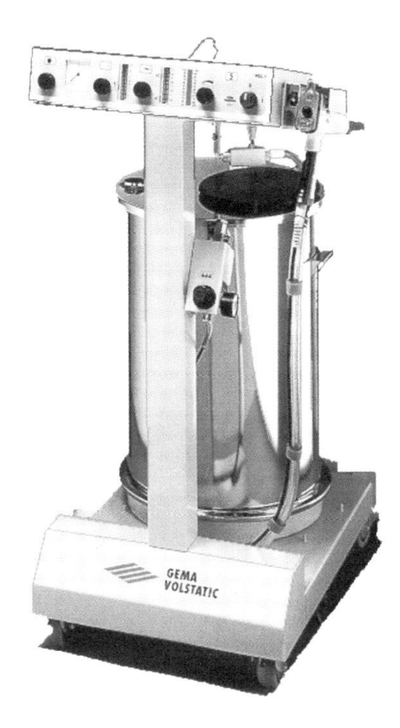

Typical manual spray gun configuration

Until box-feed systems became available, powder was usually handled using fluidised beds. They are still ideal for bulk users of powder, both for manual and automatic systems, and tend to use lower volumes of air to feed and transport the powder to the application device, which gives them better transfer efficiency than other methods.

When the system has to be cleaned, for example for a colour change, any remaining powder must be removed and the unit thoroughly cleaned. Fluidised beds are now designed to be topped up without stopping the process. The hoppers should always have close-fitting lids and be vented into the spray booth so that any dust produced by the fluidisation can be contained and the powder recovered.

Box units

The time and labour required in changing colours tend to be a headache with powder coating. Colour changes become very much quicker with box feed systems, however, in fact almost as easy as with wet paint.

Direct feed delivery systems take powder straight from the powder manufacturer's containers, which is particularly convenient for coaters who are expecting to apply a number of different colours during the course of the day. In addition, because powder does not have to be transferred to a hopper or other container, it is much cleaner.

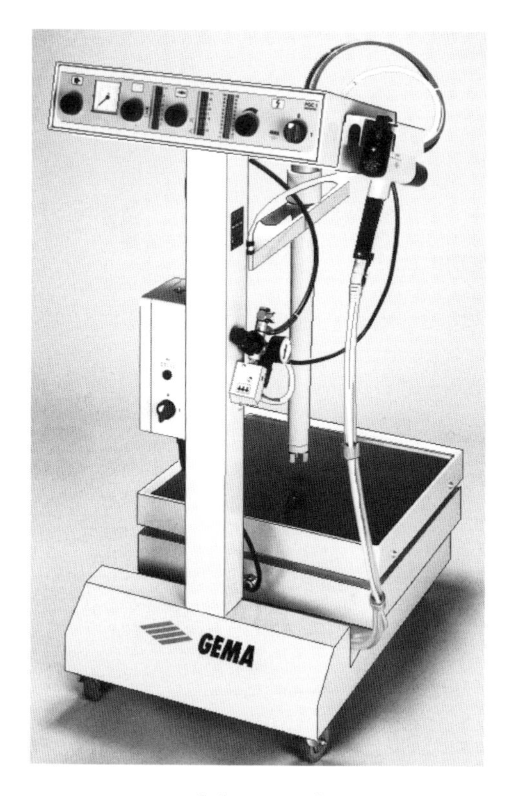

A box unit

In these delivery systems a fluidising shoe or perforated coil is lowered into the box and this provides fluidisation in the immediate vicinity using compressed air. In some systems a vibrator is provided to assist the fluidisation.

The powder feed is not as uniform as in the fluidised bed method and a higher volume of air may be required to transport it, reducing transfer efficiency. The performance of this type of unit depends on the nature of the powder used, whereas a fluidising hopper is equally at home with all types of coating powder.

Injectors and venturis

Powder is taken from the fluidising hopper or box feed system via a venturi fitted above the powder container.

The delivery of powder to the gun is critical. It is important to appreciate how powder injectors work and their vital role in the powder coating process.

To obtain even coatings and maximum transfer efficiency, the required volume of powder must be delivered to the application device free from surging and puffing.

The injector is operated using compressed air. Its purpose is to draw powder from the container, either directly or via a fluidised bed, to supply a uniform powder-air mixture to the spray system.

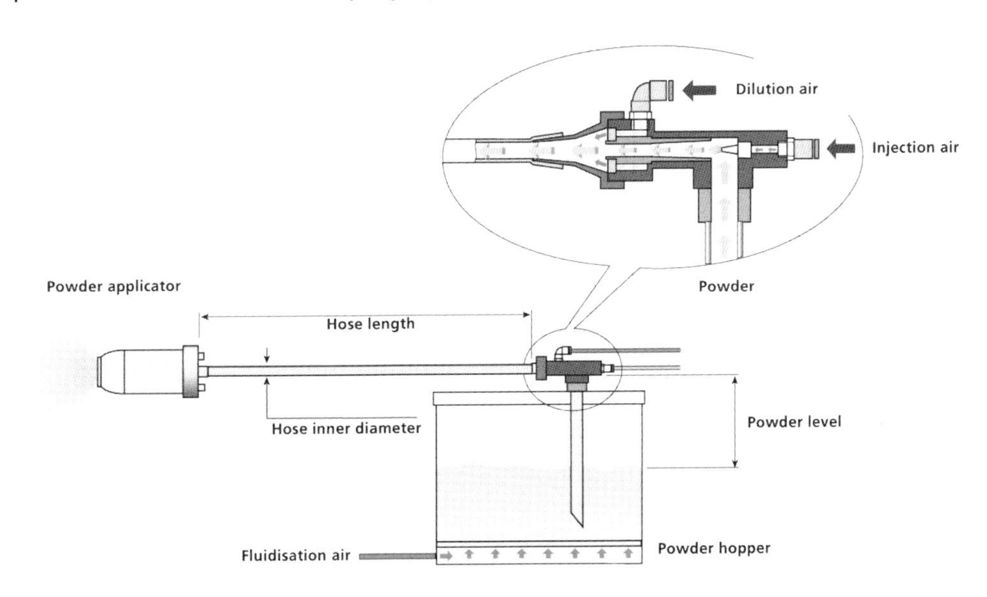

A venturi device

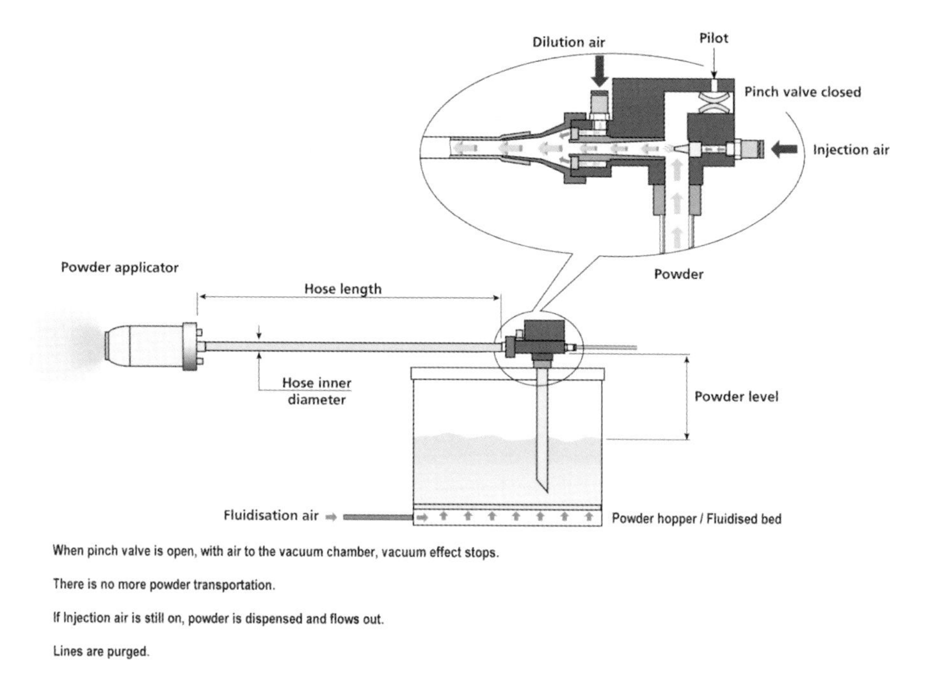

A venturi device with powder purge facility

The air projected from the injector creates an area of low pressure in the chamber above the powder supply tube and this sucks up powder from the fluidised bed or box. Its inertia propels it forward into the delivery tube towards the application device.

For effective powder application, powder must be supplied continuously at a rate between 60 and 500 g per gun per minute. Strict control is important, and anything more than the amount strictly required is waste. Less than the amount required produces too thin a film and slows down the application.

The powder to air ratio must be adjustable, for this reason most venturis now have an additional air supply located in the working part of the injector system. This is described as *dosage* dilution or *transport* air, and helps to provide an even distribution of powder.

The injector must be able to guarantee a smooth powder supply even if the system has to be started and stopped frequently. A powder purge facility will help this.

Powders are generally much more abrasive than might be expected. To cater for this the injector assembly is either coated with a sacrificial lining designed to wear evenly and be replaced, or it is made from a wear-resistant, low friction material such as PTFE. Because powders are so abrasive, all parts have to be easily replaceable in service.

The length and diameter of the tube conveying powder and air to the gun is also fairly critical. The supplier normally provides the recommended values, but by experimenting with the length and diameter of the hose it may in fact be possible to reduce the volume of air needed. Equipment manufacturers are sometimes able to offer a graph showing the ratio of diameter to length suitable for a given powder output. For an example ot this please turn to page 200 in Appendix II.

Electrostatic powder spray guns

The final element in the chain is the electrostatic powder spray gun. This is ideally light, flexible and easy to use and service, and should come with a full range of accessories.

The gun has to be capable of applying the full range of powders in use and versatile enough to over-coat existing coatings and to spray both thin and thick films.

Guns can provide either a negative or a positive charge. Remember that epoxies and polyesters are naturally negative whereas nylons are naturally positive.

Performance is directly related to the type and effectiveness of the electro-static charging method. This assumes that the powder feed to the gun is even, and that the volume of air is the minimum needed to deliver the right amount of powder.

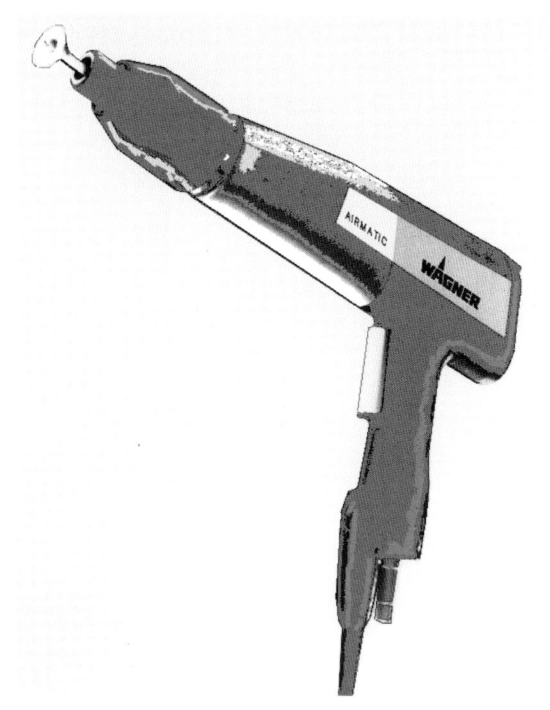

An electrostatic powder spray gun - Corona

For corona charging, with or without ion stripping, the variables are as follows:

- *Power output*. Guns with limited power will be slower and may be unable to cope with complex shapes.

- *Voltage*. It is useful to be in a position to spray at either constant voltage or constant amperage, each having its own virtues in terms of speed, efficiency and quality. Higher voltage enable faster application rates on metallic components but the voltage has to be reduced when over-coating. Similarly, reducing the voltage is better for coating into recesses and over sharp corners.

- *Current*. Maximum microamperes are best for speed, but combined with high voltage are liable to cause back ionisation. Reduced microamperes are used for recesses and for over-coating.

U max	Parts with cavity	Flat & round parts
85 kV	• Improves the powder coating of the cavities	• Improves the charge of the particles • Better wraparound • Better transfer efficiency
50 kV	• Overcoating • High thickness (+ 100 μm) • Low conductivity works parts	

5 μA　　5 μA　　80 μA　　1 max

Graph voltage and current characteristics as applied to type of coating

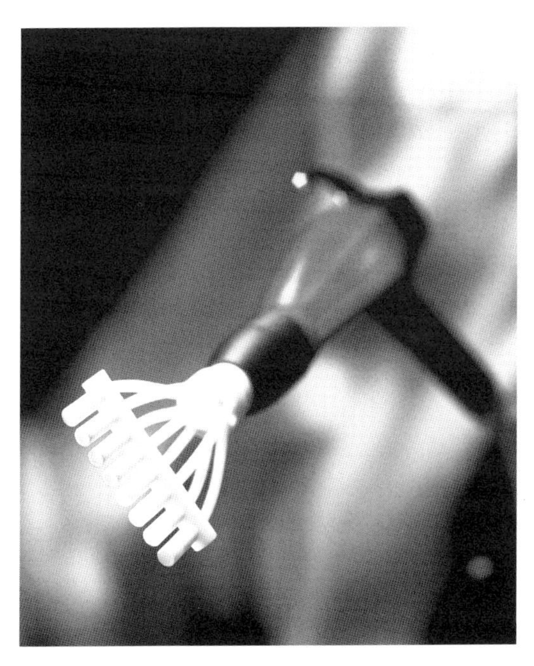

A manual tribo charging powder gun

Tribocharging guns are a very different subject. Often powders are specially designed for tribocharging and the powder manufacturer will need to know which technique you are using if he is to supply the correct powder.

This charging method depends entirely on the powder and the flow of air, which gives fewer variables to play with. Increasing the air to obtain a higher charge may reduce transfer efficiency, and slowing down or speeding up the application process will change the thickness of the coating. Best results are obtained by trial and error.

It is naturally a great help if the user can return easily to settings used previously. These may be readings of actual voltage and microamperes, or based on an arbitrary measurement of output.

Other factors

- *Weight*. The heavier the gun the more arduous the task becomes. Covering large or complex items requires greater manipulation when spraying manually. The lighter the gun, the easier it will be to use over long periods and sustain the quality of coating on complex shapes.

- *Nozzles*. These are available in various configurations. Transfer efficiency is often a function of the nozzle used for a particular component. Both fan-spray and round-spray nozzles are available, the latter using a vortex effect to give a swirling type of cloud with low forward velocity. Extension nozzles are available to coat the insides of tubes such as those used on fire extinguishers.

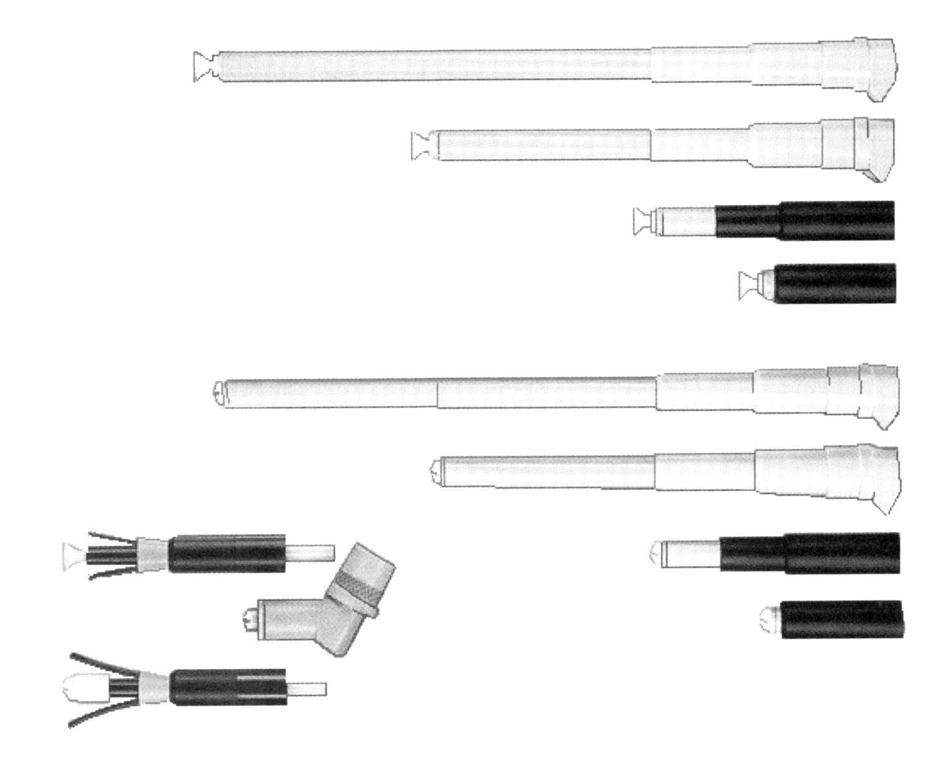

Nozzles and extensions

- *Accessories*. The wider the range of accessories the more flexible will be the range of work that can be undertaken. A high-voltage test meter, for example, is often to check that equipment is performing correctly and in accordance with the manufacturer's recommendations. Special devices are also available to withdraw provide a small sampling system for the testing of powders.

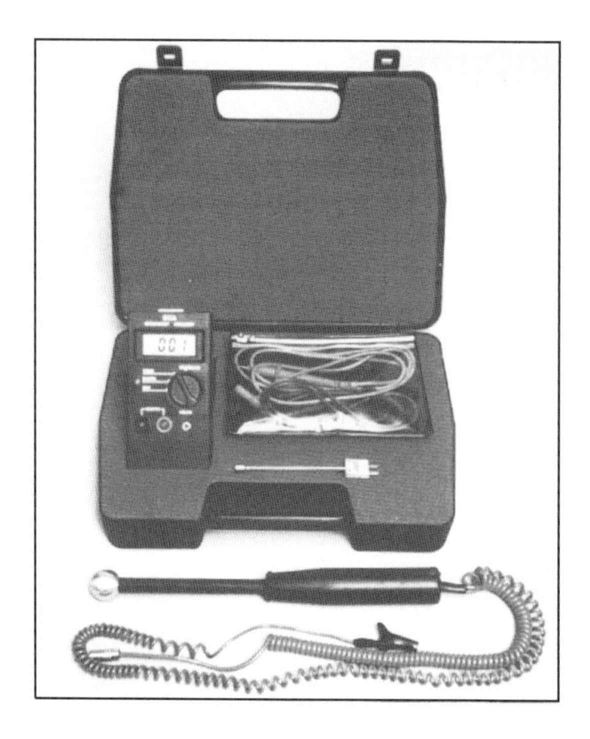

A high-voltage test kit

- *Cost.* Manual guns are relatively low in cost. Units with wider flexibility and comprehensive control will of course be dearer.

  Capital and operational costs are clearly important factors. On the other hand, cutting back on the specification of the spray gun is going to be false economy if the transfer efficiency and versatility of the unit are put at risk.

  Transfer efficiency is defined as the amount of powder sprayed relative to the amount actually deposited on the component. The hopper can first be weighed, then a number of components weighed before and after coating so that the percentage transfer efficiency can be worked out. This is a simple procedure and quite easy to carry out routinely during production.

- *Cleaning.* Manual guns are easy to clean, normally requiring no more than a blast of air through the gun plus simple cleaning of the nozzle.

When a great number of colours are to be applied manual guns offer exceptional flexibility.

- *Setting up*. Manual guns are quite easy to set up, monitor and control. Most have simple, accurate set-up procedures that can easily be adapted to meet changing application conditions.

- *Flexibility*. Manual guns offer a lot of flexibility and with the right choice of deflectors and nozzles they can be used to coat a wide range of components regardless of size and shape. Manual guns are best for small or medium production runs, both in batch and conveyorised systems. It is however important to take into account the speed of the conveyor system. Transfer efficiency, speed, user-friendliness and weight are the key points in assessing the flexibility of a spray gun.

To summarise, when choosing a spray gun the main factors are:

- Weight, assessed at arm height with hoses and cables connected.

- Transfer efficiency and speed for each particular nozzle.

- Instrumentation and repeatability of settings.

- Feel - is it comfortable to use?

CONVEYORISED ELECTROSTATIC FLUIDISED BED COATING

Electostatic powder coating can be combined with the fluidised bed process, removing the need for the pre-heat cycle. Proprietary types of electrostatic fluidised beds are very specialised and are generally manufactured for specific applications. They may be used manually but are normally designed to be automatic.

Their use has declined recently due to the fear that coating powders in certain concentrations with air are likely to be explosive and as a result the design and use of these devices has to be very carefully regulated.

The components must be relatively small as they are coated by electrostatic attraction rather than by the action of dipping them into the powder. An overall depth of about 5 or 6 cm is about the limit for a component if a uniform coating is to be obtained.

The process is good for coating screws, small brackets and other metal components manufactured in bulk.

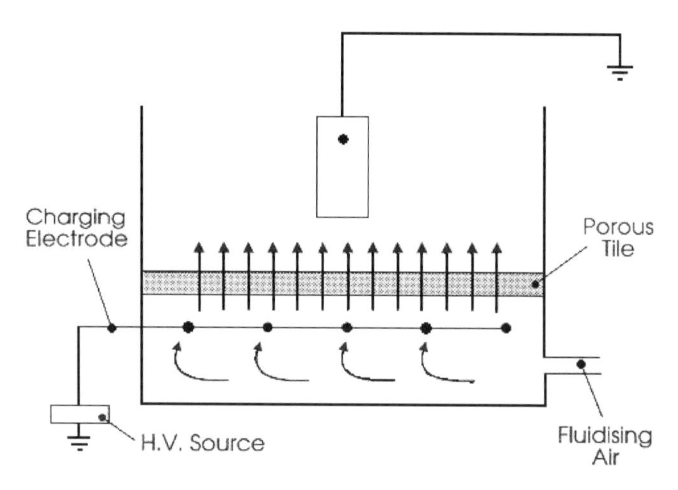

An electrostatic fluidised bed

FLOCK SPRAYING - WITH ELECTROSTATIC SPRAY GUNS

This process can be of benefit in two completely different applications. The method can be either electrostatic or non-electrostatic, as most of the coating build is obtained from the stickiness of the wet, melted surface but a charge on the coating material can be useful.

In principle, flock spraying consists of spraying fibres on to substrates with an adhesive surface, either overall or in a pattern. A well-known example is flock wallpaper, on which the coating can also incorporate anti-condensation properties. The coating applied has a soft feel and can also be very decorative.

A completely different application, of this principle already metioned in the previous section is in the chemical industry, where components such as valves, pipes and tanks can be given excellent protection by pre-heating them and then spraying on powder. The heating and spraying cycle is repeated a number of times to obtain extremely high build, in fact coatings produced this way can be as much as 1000 microns (1 mm) thick.

AUTOMATIC ELECTROSTATIC SPRAYING EQUIPMENT

The automation of the electrostatic powder spraying process seems at first sight quite straightforward, much more so than the automatic spraying of paints, lacquers and varnishes. Powder is attracted on to the component and stays there, whereas liquid paint will run and sag if too much is put on.

To obtain the best possible quality from powder coating and to obtain a uniform coverage over the whole component, however, is much harder, and this is particularly so if a precise range of film thickness has to be maintained.

If it was only a matter of spraying as much powder as possible on to the substrate this would of course be a simple matter. It would certainly give coverage but the result would be far from even. The coating might exhibit back ionisation at the edges, and excessive thickness in other areas could even give reduced flexibility and resistance to impact.

In practice the coating market taken as a whole is looking for a high standard of evenness and control of film weight, and this is the challenge facing anyone operating an automated set-up. The white-goods market and similar high-volume applications are examples where this is consistently achieved at reasonable cost.

The key factors are these:

- Handling of the components by good jigging, giving correct spacing between components and the proper presentation of all the coating surfaces to the application devices.
- Controlled flow of powder to the spray gun.
- Using spray guns with good transfer efficiency.
- Control of the surroundings in which spraying takes place, including air flow and possible interference from electrostatically conductive parts such as the frame of the spray booth.

The way the automatic spray gun is operated is crucial to the process. A number of basic types of equipment are available.

Air-assisted spray guns – corona charging

These are similar to manual spray guns, but of course without the handle.

They may be of similar size or larger than manual guns and are capable of handling higher flows of coating powder, though it is much better to increase the number of guns rather than rely on stepping up the powder flow if high transfer efficiency is to be achieved.

Corona units, with and without ion stripping devices, can be used. Depending on the manufacturer, the high-voltage source may be internal or external to the spray gun.

At one time voltages up to 150 kV and currents as high as 200 microamperes were not uncommon. More recently concern about the hazards involved has led to the use of lower voltages and currents, and this is now normal practice. A typical voltage might be in the range 85 to 100 kV and maximum current roughly 80 to 100 microamperes.

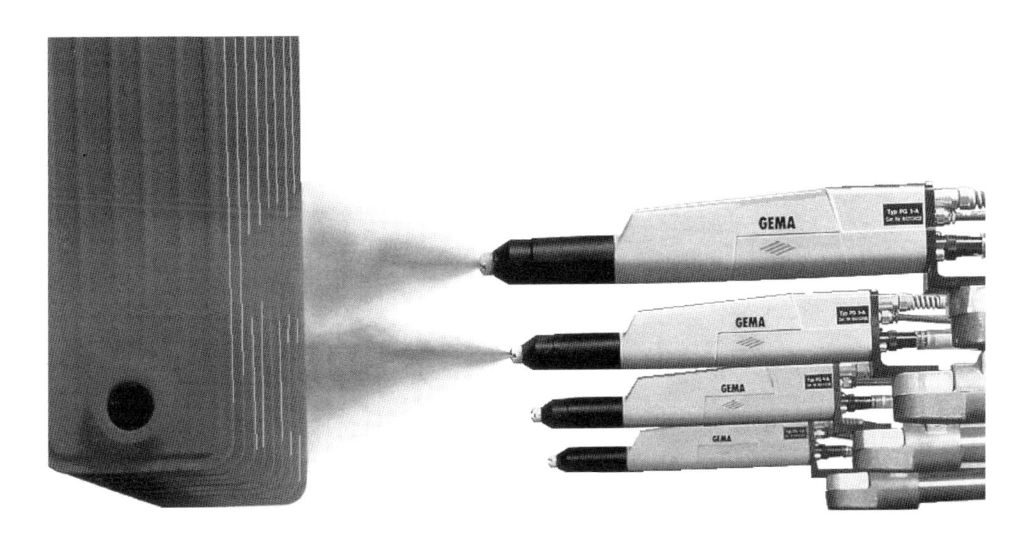

Automatic air-assisted electrostatic spray gun – corona charging

The powder maximum output of a corona charging gun can be up to 500 g a minute or 30 kg an hour. Above this point transfer efficiency declines dramatically, and should ideally be limited to half these values.

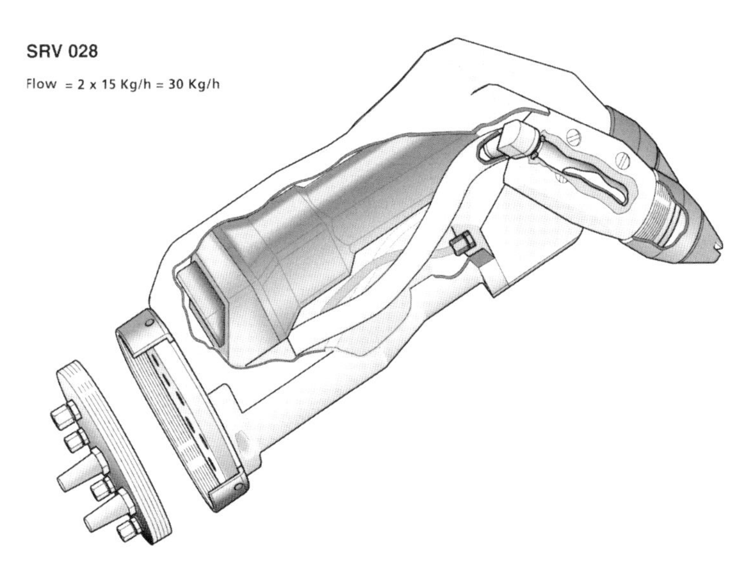

An adjustable air assisted unit for complex component shapes

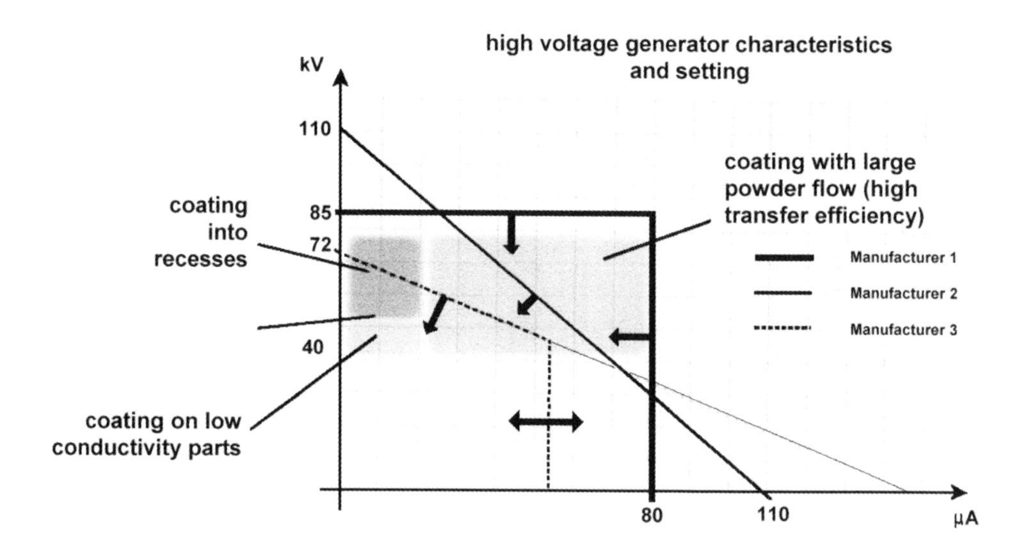

The above graph also shows the settings for different types of coating:

- For maximum transfer efficiency of the powder: use maximum voltage and current but watch for back ionisation on edges and thicker areas.

- For over-coating or re-coating of components: use slightly reduced voltage but greatly reduced current.

- For coating substrates of low conductivity such as MDF or glass: these are not easy, and both the voltage and current must be reduced for good results.

Automatic air-assisted electrostatic spray atomisers – tribocharging

Tribocharged automatic spray guns can be very useful for powder coating as they give more even film weight, with better penetration into enclosed areas and improved control of film thickness.

In general, a higher number of tribocharged spray guns will be needed to give the same application speed as corona systems. For good transfer efficiency the powder flow should not exceed 10 kg per hour.

Powder bells

These devices are similar to those used for liquid paint. The bell-shaped atomising head rotates at variable speeds up to 10,000 rpm using an air turbine. Increasing the air surrounding the bell changes the shape and forward momentum of the powder to suit the application. With low powder output, say 250 g per minute, a rotating bell applicator can achieve transfer efficiency greater than 75%. It is possible to obtain a similar effect using a stationary bell arrangement. This is similar in principle to an air-assisted spray gun but with a wider spread of powder and reduced transfer efficiency.

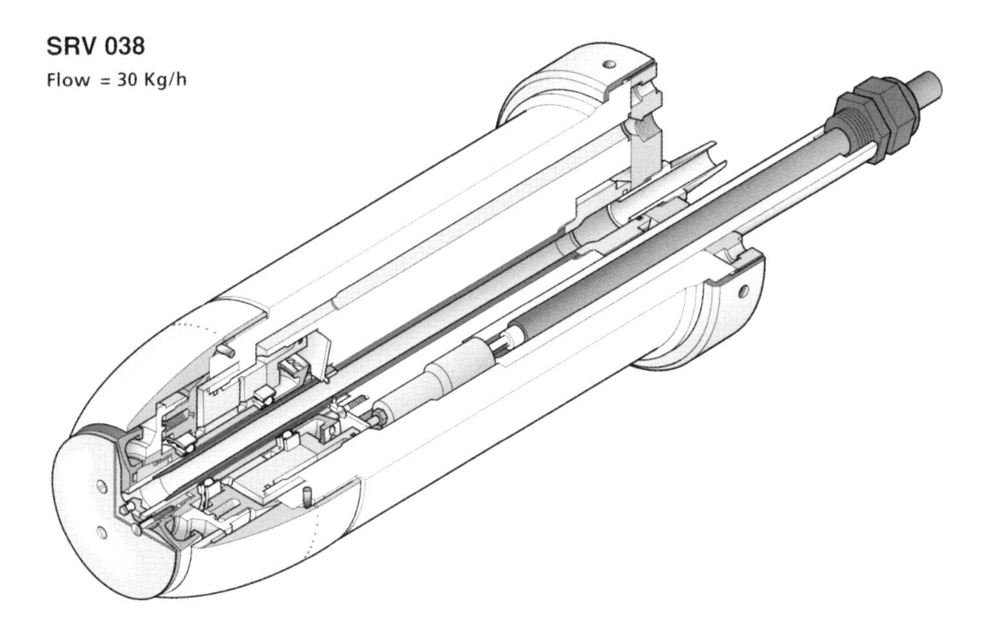

A typical powder bell

Powder discs

The powder disc can similarly have a stationary or rotating final atomising and distribution disc. These units are used for horizontal spraying and move in an upward or downward mode. They are particularly useful for spraying both the insides and outsides of cylindrical components.

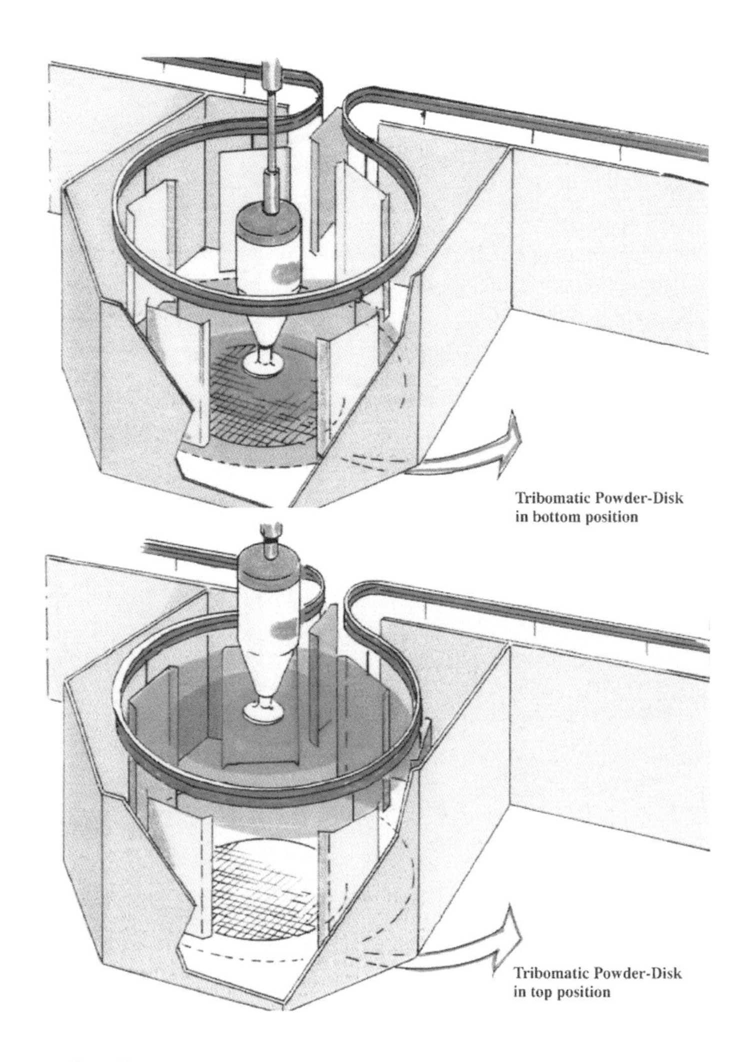

A tribo disc in an 'omega' loop configuration

They are also used in the omega-loop conveyor configuration that is so effective for liquid paint and varnish application, the applicator moving within the conveyor loop.

These units are generally tribocharging applicators and have several tribo devices within the shroud of the unit. They feed powder into a distribution head, giving a virtually horizontal cloud of powder. Corona charging discs are similarly used.

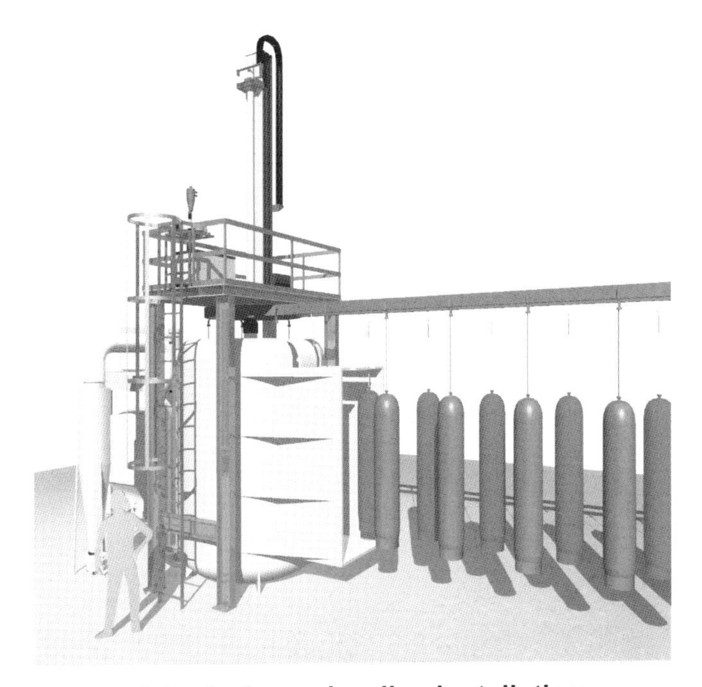

A typical powder disc installation

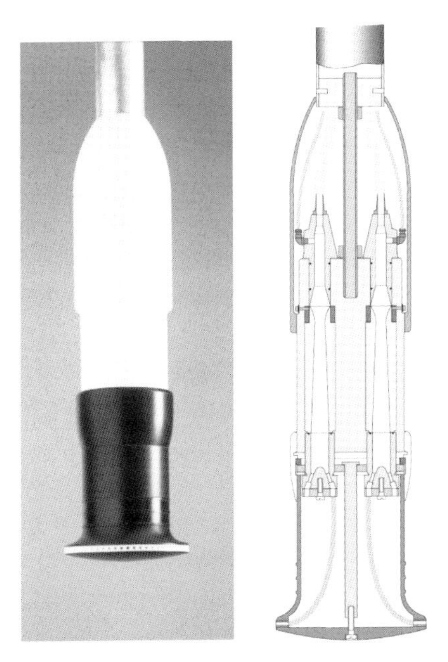

A typical tribo disk powder applicator

AUTOMATIC APPLICATORS / MANIPULATORS

In the simplest case the components are held in a fixed position and the applicator moves around them. A number of devices are available to move the automatic spray applicator. These may merely reciprocate up and down vertically or the speed may be variable with the ability to dwell in particular sectors of the movement, and possibly move along a second or third axis.

Fully programmable robots give the ultimate flexibility and control but they need to be able to cope with dusty conditions.

The choice of automatic spray applicator will in the end depend on the particular application involved and the selection confirmed in trials carried out under conditions similar to those expected in practice. Most suppliers of powder coating systems have laboratories equipped with the full range of automatic equipment where demonstrations can be arranged.

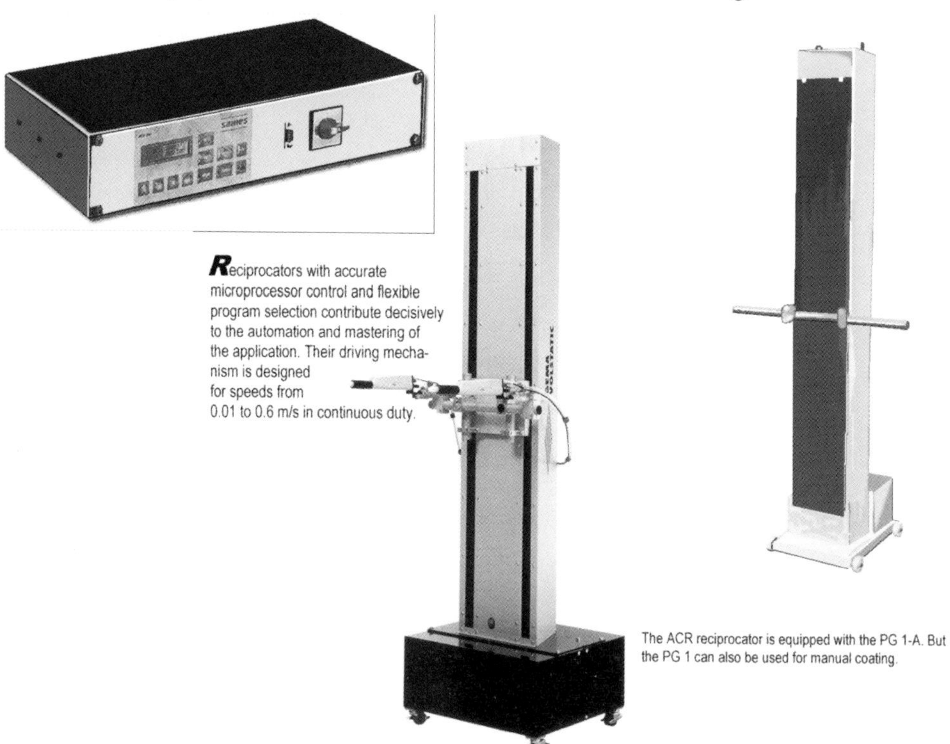

Reciprocators with accurate microprocessor control and flexible program selection contribute decisively to the automation and mastering of the application. Their driving mechanism is designed for speeds from 0.01 to 0.6 m/s in continuous duty.

The ACR reciprocator is equipped with the PG 1-A. But the PG 1 can also be used for manual coating.

Applicator manipulators

It is best to carry out these tests using the components to be coated and the actual powder that will be used. All variables should be recorded and samples kept for reference in the future. Good photographs are an ideal way of remembering how applicators were configured and a video is even better.

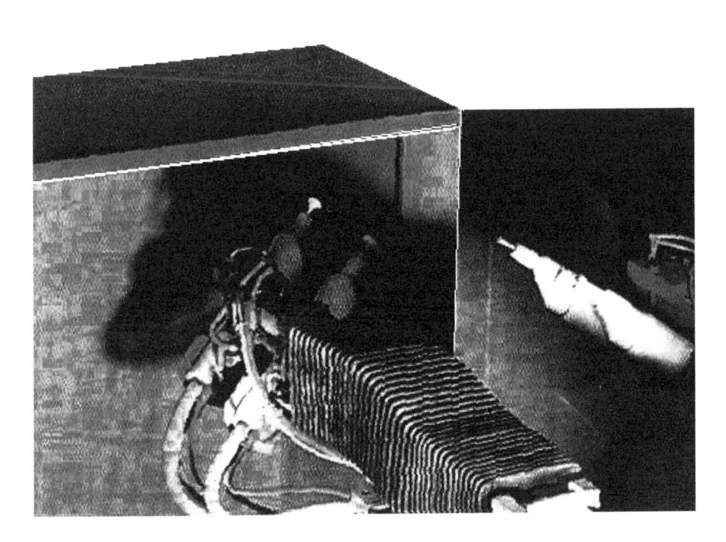

A powder coating robot system

CHAPTER V

POWDER SPRAY BOOTHS

Chapter V

POWDER SPRAY BOOTHS

The purpose of the powder booth in electrostatic spraying is to ensure that powder from the application device stays inside the booth. This then offers the opportunity to collect the overspray and recover it re-use.

HEALTH AND SAFETY

Dust is obviously inevitable with powder coating and it must be controlled well enough to protect operatives from inhaling it. Additionally, conditions have to be avoided in which the concentration of dust in the air might become an explosion or fire hazard.

Coating powder, being an organic material of small particle size, becomes an explosion hazard if its concentration in air is above the *lower explosive limit* (LEL). For most coating powders the LEL falls in the range 20 to 70 g/m^3 and to give a margin of safety most advisory and legislative bodies require that the powder concentration in air in any equipment handling coating powder should not exceed 10 g/m^3.

It is the duty of plant designers to ensure that sufficient air movement is provided to accommodate in safety the amount of powder being handled at any particular time. A reputable supplier will stipulate the polymer type, the particle size distribution, and values for the maximum output of the application equipment and the air movement within the powder handling plant.

Explosions might also occur when hot surfaces are present or if there are discharges of static, and the plant design must allow for this. Ovens should be installed well away from the application area, all electrical equipment must be adequately earthed and the application equipment itself should meet the standards set by the national authority.

Other safety issues include good housekeeping and the need to keep plant free from powder build-up, and also ensuring that naked flames and smoking are not allowed near the plant.

The components to be coated should be properly earthed, as static charges can build up when they come under the influence of the application device. A resistance of less than 10^6 ohms is suggested, measured by the methods given in UK Standard BS 2050 or its equivalent.

In the case of manual application it is particularly important to safeguard the health and safety of the operator by making sure that air travels past him into the booth in sufficient volume to avoid contamination of the area he is working in. Any powder escaping from the booth can create a health hazard and local authorities demand regular checks to make sure this is not happening.

In the UK, the British Coatings Federation booklet *Code of safe practice: application of powder coatings by electrostatic spray* states that air flow in a manual booth should be from behind the operator, across the component and into the body of the booth. The minimum average velocity should be between 0.5 and 1.0 m/sec.

It is clear from what has been said that the spray booth is a key area of concern in safeguarding the health of operators. Powders containing toxic compounds such as triglycidyl isocyanurate (TGIC) require particular care. This material is toxic by inhalation and swallowing and is an eye and skin irritant. Many authorities set exposure limits; in the UK it is currently 0.1 mg/m^3 for this substance.

Coating powders may contain other ingredients that are a threat to operators' health, and suppliers' Health and Safety Data sheets must be carefully examined and safeguards put in place as required.

MATERIALS OF CONSTRUCTION

The material used for constructing spray booths has been an issue for many years. Steel, either coated or stainless steel, is hard wearing and can be fabricated to give smooth-sided, pocket-free enclosures that are easily cleaned. The conductivity of this type of booth aids the discharge of static electricity and they can be regarded as completely safe in this respect.

Construction using plastics and composites on the other hand inevitably brings with it the problem of charge build-up, and thus the possibility of static discharge and powder explosions. Surfaces such as these of low conductivity have a benefit in one respect, however, as they repel powder, and transfer efficiency is improved as a result. Plastic surfaces can also be cleaned faster and more thoroughly.

There are several international standards giving guidance on the materials that should be used. European Standard EN 50177:1966, for example, states that plastics used in spray booths having a wall thickness less than 9 mm can give rise to static discharges, and that any materials thinner than this should have a breakdown voltage not exceeding 4 kV.

The design, materials of construction and operation of powder spray booths is a key element in maintaining high transfer efficiency.

Spray booths are generally box-shaped with an aperture through which the operators can spray manually. Alternatively they can be designed for automatic spraying, in which case there will be an aperture on one or both sides depending on the application. Normally these are staggered.

Powder spray booth operator inside

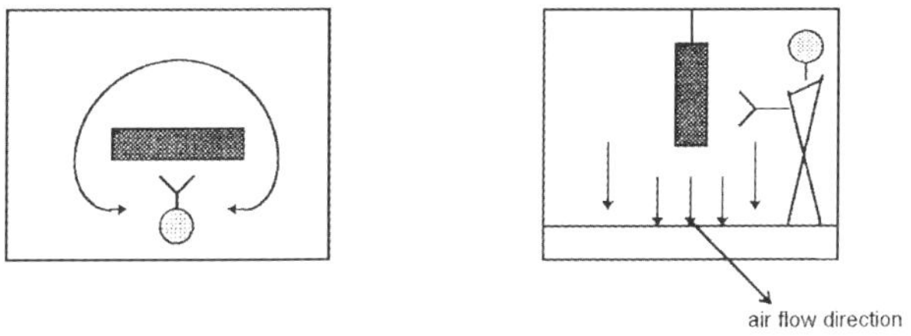

air flow direction

Powder spray booth with operator outside

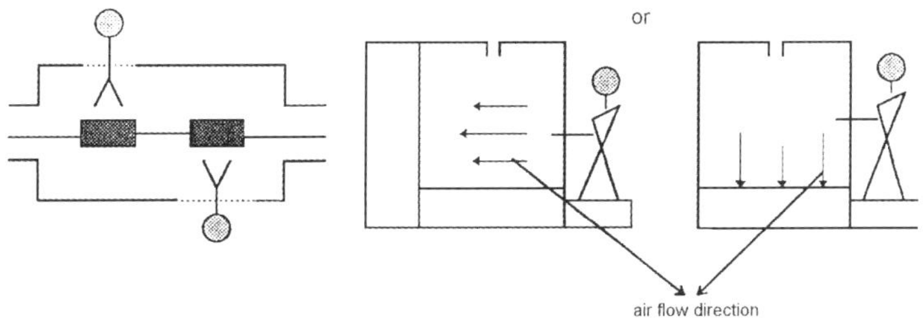

or

air flow direction

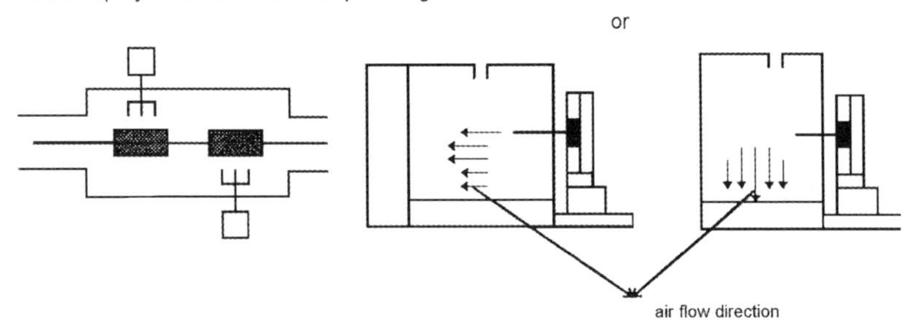

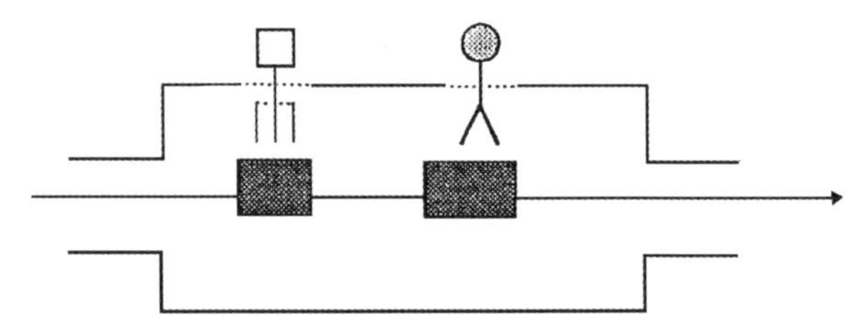

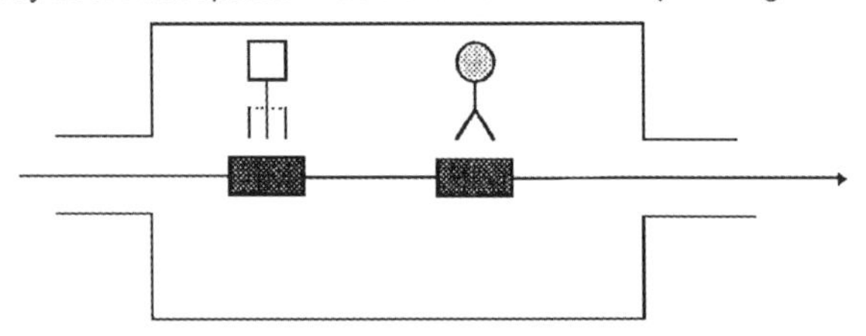

Spray booth configurations

DESIGN

As mentioned above, European Union legislation gives guidance on the design and safety requirements for 'powder spray booths for electrostatic application with organic powders'.

The components to be coated are either manually loaded into the booth or automatically transported through the entrance and exit vestibules past the spray guns.

In recent years it has become more normal for the operatives to work inside the booth, but special ventilation requirements are needed to ensure their health and safety.

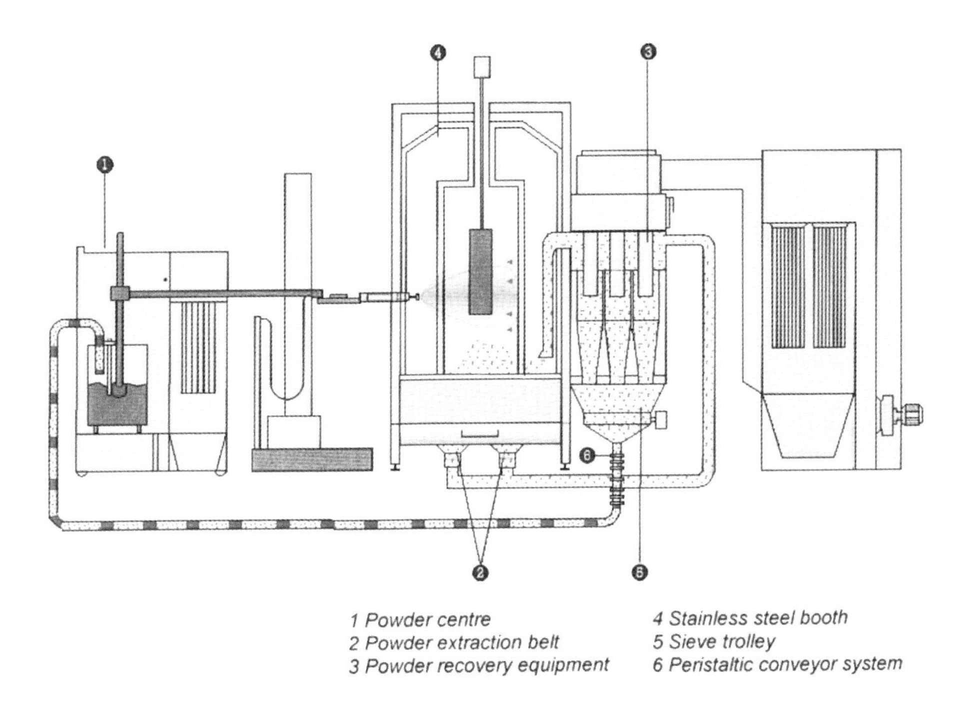

Powder flow within a plant

CLEANING

An important feature of good design is that it should permit easy and thorough cleaning. The cleaning technique recommended must be used and it is preferable to use equipment such as a vacuum cleaner and squeegees rather than high-pressure blow-guns. These can be dangerous as they are prone to re-charge the powder, so that it becomes even more tenacious and difficult to remove completely from the walls. When this happens it can cause it to become contaminated with powder previously used.

The velocity of the powder-air mixture in the ductwork should be sufficient to make the duct self-cleaning. Air velocities greater than 10 m/sec will ensure that this is the case as long as the ductwork, like the interior of the spray booth, is smooth and free from cracks and corners in which powder might collect.

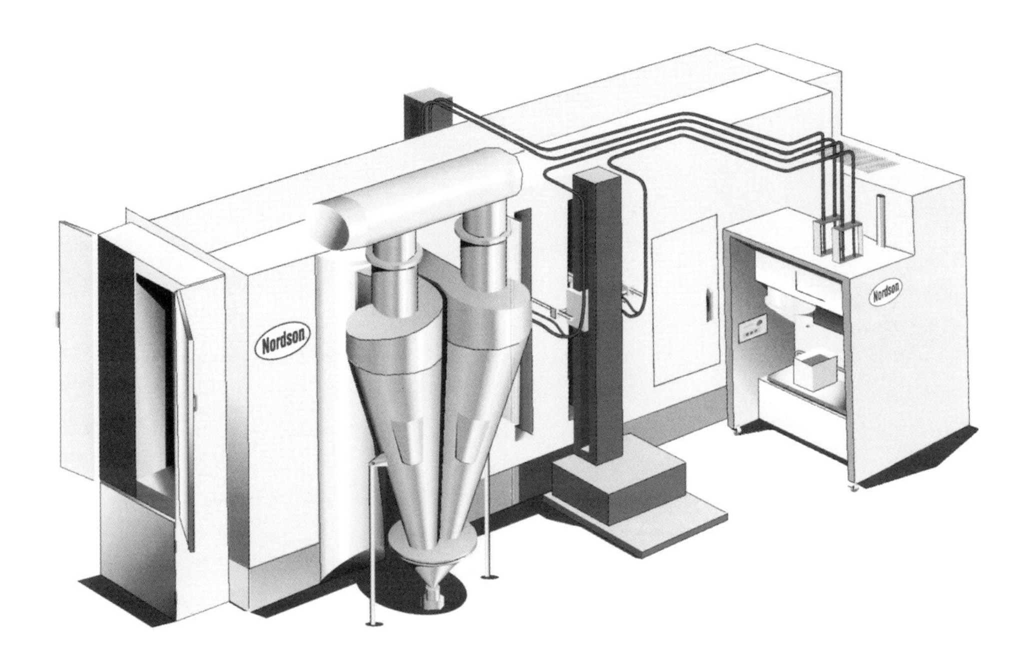

A typical electrostatic powder coating plant

SELECTION OF SPRAY BOOTH TYPE

There are many types of powder booths. Different designers have their own ideas and users often have strong individual preferences. The factors affecting the suitability of a booth for a particular type of powder coating work are, however, relatively straightforward:

- Whether manual or automatic application is to be used, or both.

- Optimum transfer efficiency of the coating powder.

- Shape and size of the components.
- Colour change requirements.
- The factory space available.
- Cost limitations.

Manual booths

The major features in this case are:

- Component size.
- Application speed and the number of spray operators.
- The speed of colour change required.

There are, of course, a great many competing booth designs from different manufacturers, all with their own advantages and disadvantages. Cleaning down and colour change requirements are the major factors in deciding which type of booth to choose. Some are best for the application of single or a small number of colours only, and others are better for companies handling a wide colour range.

Beyond this, the ability to contain the powder should be a high priority for the designer and user alike. Many booths are in practice disappointing in this vital requirement and give rise to coating problems, contamination and rejects.

Novel designs are seen from time to time and over the years some have become established. Unfortunately others are less effective in practice than their designers would suggest and fail to come up to expectations.

A frequent problem is that the operator is not able to reach the components on a conveyorised system. Some of these may be as much as 1 metre wide and 1.5 metres high, and it can be impossible to apply the coating evenly without having to reach so far into the booth that he is working within the charged powder cloud. This is a serious health risk and is clearly unacceptable.

Booth designs with movable walls are helpful, and an even more radical solution is to design the booth so that it is safe for the operator to work right inside it. In this case the air movement is downwards, the powder being exhausted through the floor of the booth. This is an ideal solution for spraying very large items such as earth-moving equipment, where the operator has the freedom to move around the component during spraying.

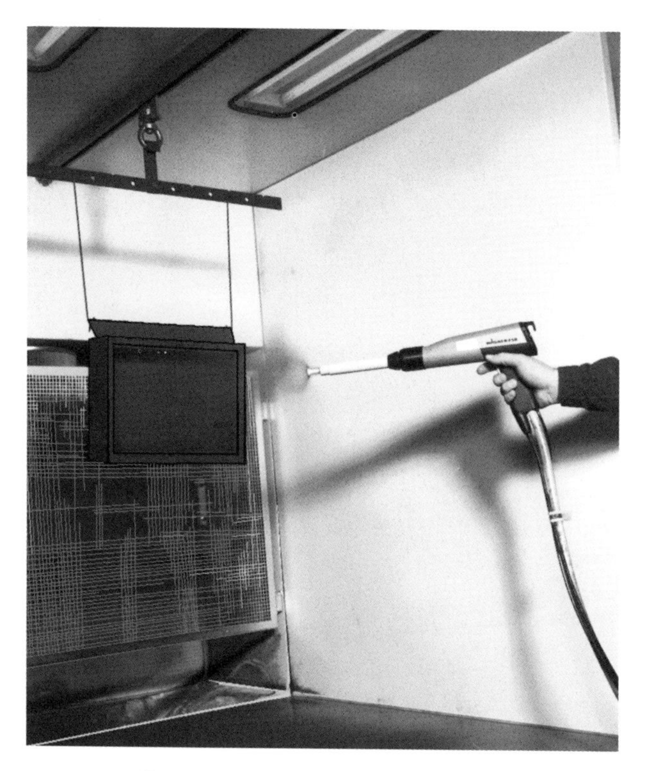

Good spray operator access

The design factors for a manual booth are as follows:

- *Adequate air flow*. This must avoid the chance of an operator breathing in powder by directing the air past him into the booth, and if he is working actually inside the booth, vertically downwards. In the first case the air flow should be at least 0.4 m/s or preferably higher, and in the second at least 0.3 m/s. There should also be no local dead spots.

- *Noise level.* Spray booths can be unpleasantly noisy and levels should be within the legislative requirements, certainly below 82 dBA and preferably as low as 75 dBA.

- *Lighting.* Illumination must also be adequate and ideally above 600 lux at the point where the spray operative is working. The quality of spraying is very much governed by the sprayer being able to judge the coverage he is achieving, and good lighting conditions are essential.

Spray room and sprayer

- *Electrical safety.* Equipment such as spray guns and extractor motors should be linked for safety reasons, so that spraying cannot take place unless the exhaust system is operating at the required level. There is also the likelihood that at some time in the future booths may have to be fitted with both fire detection equipment and devices for monitoring effective earthing. In the latter case, poor earthing of a component would automatically shut down the spraying operation.

- *Operator safety*. Operators must use respirators and protective clothing to prevent powder being inhaled or making contact with the skin. Importantly, the operator should himself be properly earthed by using conductive footwear. In the case of operators working inside the spray booth the escape route has to be good enough to allow them to escape quickly in emergency.

Notices need to be displayed giving warnings, mandatory requirements and general advice to comply with health and safety legislation.

Automatic booths

In this case the main factors affecting the design are:

- Size and shape of the components.

- Application speed and the number of guns to be used.

- Manual touch-up requirements and hence the number of spray operators.

- The number of colours and polymer types expected, and the speed of colour change required.

Automatic powder spray booths tend to be very specialised and are often unique to their designer and manufacturer. Major suppliers are able to offer interesting and sometimes patented features intended to improve transfer efficiency and clean-down time.

Selection and choice are not always easy and It is advisable to analyse the end-user's requirements carefully and arrive at a cost-benefit analysis for each element in the coating process. Fortunately it is possible to arrange for demonstrations and trials to help decide how many guns are needed, what arrangement are best for handling the components, how much manual touch-up should be allowed for, and so on.

Booth sizes now vary from powder tunnels in which very small components are coated to those big enough to powder coat a car body. In all cases the same general principles apply.

The booth must obviously be large enough to accommodate the component and the spray guns. Other requirements for automatic spray booths include the following:

- Efficient transport of oversprayed powder into the recovery unit without involving high velocities of air within the spraying area.

- Air flow must meet the legal and practical requirement to retain powder inside the booth, as well as keeping in mind the lower explosion limit of the powder.

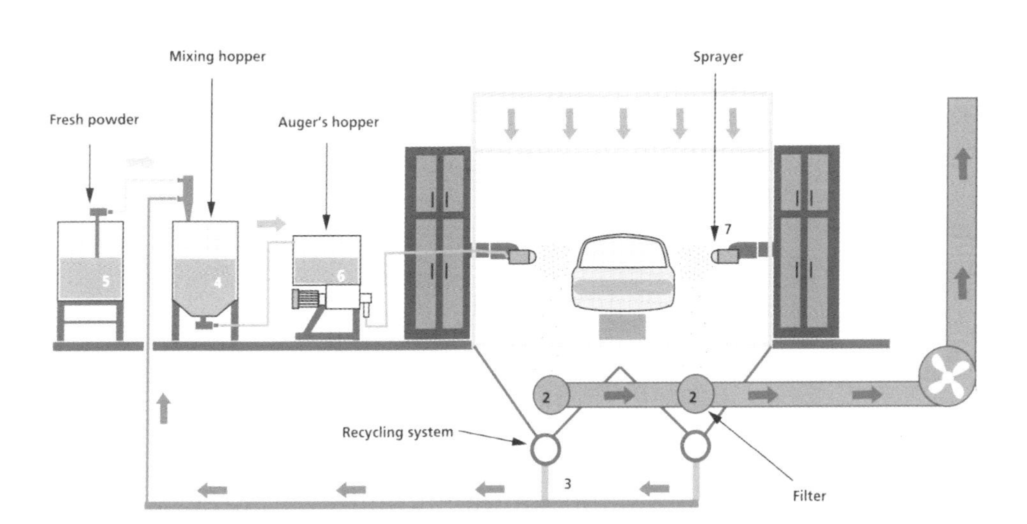

General schematic of an automatic spray booth

- Interlocking the ventilation system and the application devices, plus similar control of the doors used for access by personnel performing cleaning and maintenance.

- Ensuring that the application devices and static cables are a proper distance from conductive parts of the installation.

- Earthing of the components being coated.

- Provision of explosion relief doors and monitoring powder-air concentrations. This is even more important than for manual booths since the quantity of powder sprayed at one time will be much greater.

- Automatic fire extinguishing systems in powder collection systems where the electrostatic energy can exceed 5 mJ.

- Adequate earthing of the component to the conveyor. A maximum resistance of 1 ohm is recommended.

An automatic robotic powder coating system

COLOUR CHANGE

One of the main disadvantages of powder coating to many potential users is the difficulty of coping with colour change. There will always be pressure to speed up colour changes and this is probably the biggest challenge for the designers of both manual and automated systems.

Paint users have always had the advantage of being able to change colour merely by washing out with solvent, waste solvent and oversprayed paint being collected into waste disposal systems of one type or another. Computer-controlled systems in automatic painting plants can perform programmed colour changes in seconds, as for example happens between car bodies in a modern automotive manufacturing plant. Powder coating can never be as easy, as overspray is hard to get rid of and hangs around to contaminate the next colour.

There are a number of ways in which production and marketing colleagues can co-operate to reduce the problem. For example:

- Might it be cost-effective to hold bigger stocks of finished goods in order to make colour runs longer?

- If this is not possible, would it perhaps be worth spending money on the capital investment needed to solve the problem?

- Could it be economic to allow powder to be sprayed to waste between colours?

- Is there any chance of reducing the number of colours required?

In practical terms, contamination can arise from any air movement at all in the production area, even from people walking around. The problem becomes worse if there is not enough space in the conveyor system for colour changes to be comfortably handled.

The cost of solving the problem in the production area has been referred to above. The alternatives are as follows.

A bank of spare booths could be provided so that they can be taken out of use for cleaning. There can then be more than one booth reserved for each colour, the minimum requirement being for one booth to be in use while one is off-line being cleaned. They must be capable of being cleaned rapidly without creating further contamination.

Another way would be to install a 'power and free' conveyor system with a number of booths on different legs of the conveyor, and perhaps one booth kept for small runs with powder being sprayed to waste. It is important to note that different colours cannot be sprayed at the same time as air movement within the oven can create cross-contamination. The problem can be avoided by using infrared panels to melt the powder before the final curing stage.

To avoid the capital cost and space involved in adding extra spray booths, it is now possible to clean down and change colour within a few minutes by providing disposable plastic film on the walls. This can be rolled up and removed between colours. The fastest time is around 10-15 minutes.

It also helps if booths are constructed with smooth interiors and tunnel-like design that allows powder to fall easily into the recovery system. This can be achieved using plastics in composite form and can be designed so that oversprayed powder is electrostatically repelled by the booth's surfaces.

Systems have been designed using automated squeegees to clean the booth, and vacuum devices and similar ideas to keep the booth free of powder so that cleaning between colours is minimal.

Some proprietary systems use an automatic water wash to flush out the powder followed by hot air to dry the booth ready for the next colour. This is an interesting approach with a lot of promise.

With all these possibilities the powder supply systems to the application devices become more complicated, as do problems associated with powder recovery.

Cartridge booth for maximum efficiency

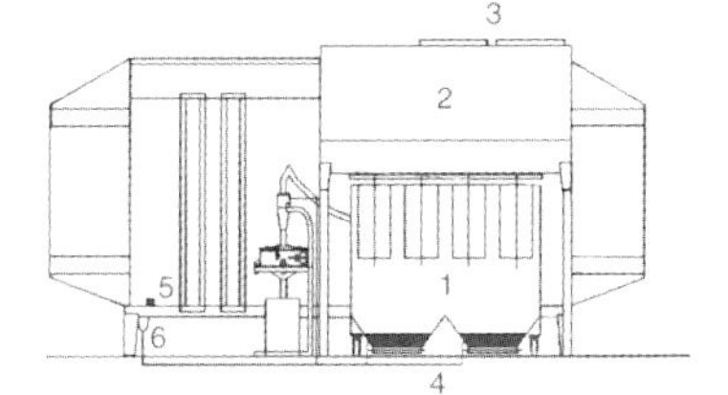

1 Cartridge filter module
2 Fan section
3 Clean air exhaust
4 Powder recycle
5 Floor sweeper
6 Floor powder recycle

Twin cyclone booth for maximum flexibility

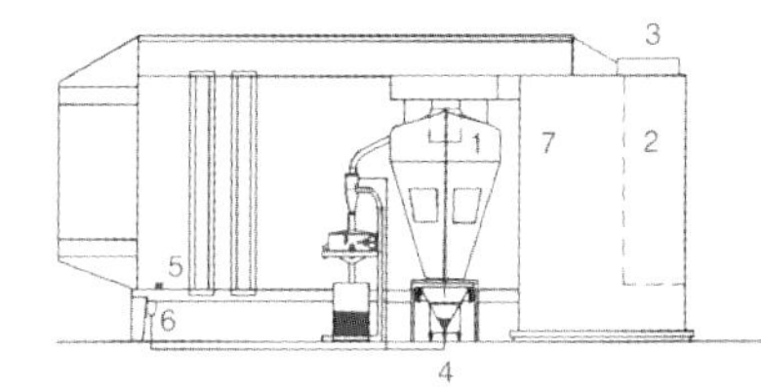

1 Twin cyclone module
2 Fan section
3 Clean air exhaust
4 Powder recycle
5 Floor sweeper
6 Floor powder recycle
7 Cartridge after filter

Multi cyclone booth for maximum efficiency and flexibility

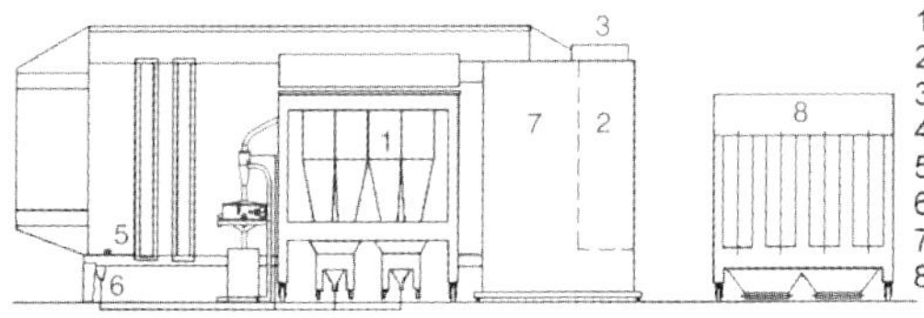

1 Multi cyclone multi colour module
2 Fan section
3 Clean air exhaust
4 Powder recycle
5 Floor sweeper
6 Floor powder recycle
7 Cartridge after filter
8 Interchangeable cartridge module

Colour change scenarios

In practical terms the colour change issue boils down to three factors:

- The space available
- Personnel numbers
- Capital cost

This can be illustrated using three actual examples:

Plant 1

Operates a 4-metre long booth with plastic film walls that can be rolled up and removed during colour changes. It is equipped with two reciprocating air-atomising spray guns on either side and a fluidised bed hopper for each colour. Cleaning and colour change is carried out by 4 men in 9 minutes. While the plant is working each man is occupied for approximately 30 to 40 minutes cleaning up and preparing for the next colour change.

Plant 2

Uses a 3-metre booth provided with a water-wash system, and an effluent plant to remove the powder from the wash water before discharge to waste. Colour changes require 2 men for 18 minutes.

Plant 3

This plant has plenty of space and capital has been available for investment. A faster and less operator-dependent system has been installed involving automated movement of spray booths on- and off-line in the coating lines. Colour changes are possible with minimal operator involvement in 6 - 8 minutes.

CHANGE IN POLYMER TYPE

In the past there have often been problems when coating powders with different base polymers are used in the same plant, for example some nylons are in compatible with epoxies and polyesters. This problem has now been largely overcome, though there are still occasional difficulties due to the incompatibility of acrylic-based powders and other polymer systems. These may be also overcome in time.

Swapping from one polymer system to another is still not always easy, however, and requires trials and careful thought. Even changing from one manufacturer's polymer to another or from pigment to pigment (e.g. between colours and metallics) can sometimes be an unexpected cause of incompatibility.

Unlike the majority of paint systems, contamination is always an inherent problem with coating powders and it is a constant source of worry to plant managers. The rule has to be: "If in doubt, don't do it" – or at least check first. Problems may not show up until after the curing process.

THE IMPORTANCE OF TRANSFER EFFICIENCY

The transfer efficiency of powder application is just as important as with solvent-based paints. In both cases the coating wasted by not hitting the component at the first attempt represents an unwanted cost in terms of materials, energy and processing. An extreme example of this is when companies decide it is worthwhile spraying to waste to save time between colours.

It might be felt that overspray of coating powder is not particularly critical as it can be recovered and re-used. Recovery and re-cycling of coating powder itself costs money, however, and the less there is to recover the better. It follows that the higher the transfer efficiency of the application equipment the lower the cost. The main reasons can be summarised as follows:

- Powder that is recovered, recycled, and reused has a greater chance of becoming contaminated. When contamination does occur it will ruin the whole batch of powder in use at the time.
- Recycling the powder alters the particle size distribution. Coarse powder tends to be recovered preferentially and fine powder is lost to the after-filter. Electrostatic application prefers the middle of the size range between 15 and 35 - 40 microns, and recovered powder has to be diluted with virgin material to retain its performance.

Recovery, recycling, sieving, and mixing with virgin powder all takes time and effort, even if it is fully automated. Manual recycling operations are labour intensive and create high levels of dust and potential contamination, not to mention how unpleasant they are for the operators.

The higher the transfer efficiency the lower the overall powder coating cost. We have seen elsewhere that it is not difficult to measure, it is simply necessary to weigh the powder feed hopper before and after applying powder to a number of pre-weighed components. These are then weighed after coating and curing. By comparing the total weight of powder sprayed to the weight of powder deposited, the percentage transfer efficiency can be worked out. This is far simpler than the corresponding calculation for liquid paint as solids content does not have to be taken into account.

Provided an operator sprays consistently, the routine monitoring of transfer efficiency is a valuable check on the effective use of powder and the equipment.

POWDER RECOVERY

It is well recognised that one of the major advantages of powder is that it can be recovered after spraying and re-used. On the other hand as we have seen above it is certainly a major concern that transfer efficiency should be as high as possible, with as little to recover as possible.

Recovery of waste powder is important and we will examine it in some detail.

The amount of air involved can be surprising. Imagine a booth with the following openings:

- Two openings of 1.5 by 1.5 m to allow components to pass in and out. Total area: 4.5 m^2.
- Two manual operator openings of 1.5 by 1 m required for touch-up. Total area: 3 m^2.
- Aperture for a manipulator handling automatic spray guns. Total area: 0.75 m^2.
- An overhead conveyor opening 3 m long, giving an additional area of 0.45 m^2.

This adds up to a total aperture area of 8.7 m^2, and if we assume that air must travel through at an average speed of 0.5 m/s, this equates to an air movement at the rate of 4.35 m^3/s or 15,660 m^3/h, not allowing for inefficiencies. A recovery system must be capable of coping with this huge level of air flow.

There are a multitude of proprietary systems for recovering powder and these fall broadly into two types:

- Cyclones
- Filters

Suppliers sometimes employ both principles together.

Cyclone recovery

The traditional system for powder recovery has been to use a single cyclone connected to the spray booth. Cyclones are not however as efficient for recovering powder as one could wish and an additional filtration step is often needed.

The final filter will have an electric fan mounted on the exit side sufficient to draw the correct amount of extract air through the system. This must be enough to cater for the booth openings, the cyclone and the restricting effect of the powder collecting in the filter bag.

The exhaust air should be clean enough to be returned to the factory and does not have to be vented to the outside. This gives economies in energy as there will be no loss of heat. This is a huge advantage over solvent-based paint application, where there is no choice but to discharge to atmosphere.

Cyclones are sometimes self-powered but in general the air is pulled through from the final extraction unit.

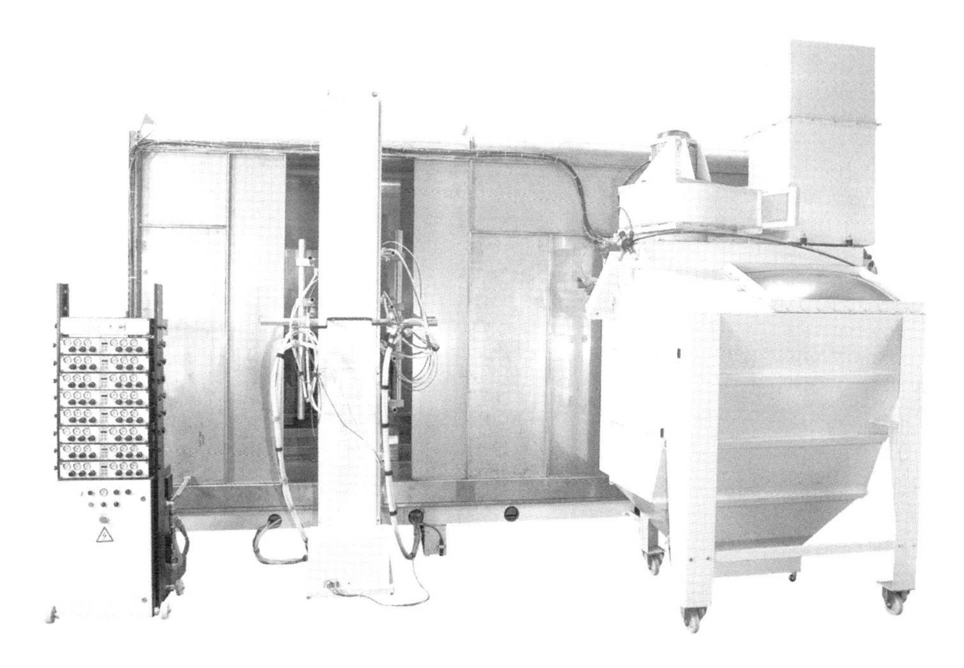

A cyclone recovery system with integrated after filter

3 way-valve

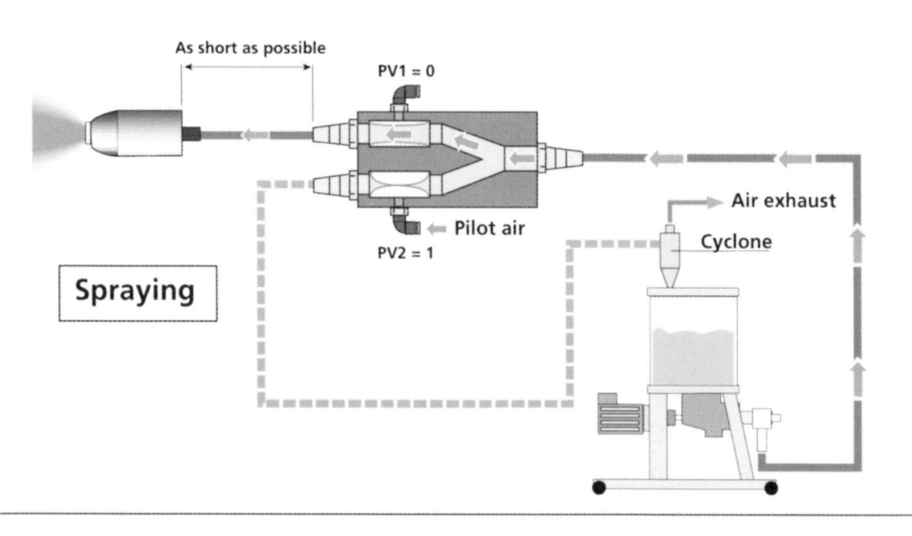

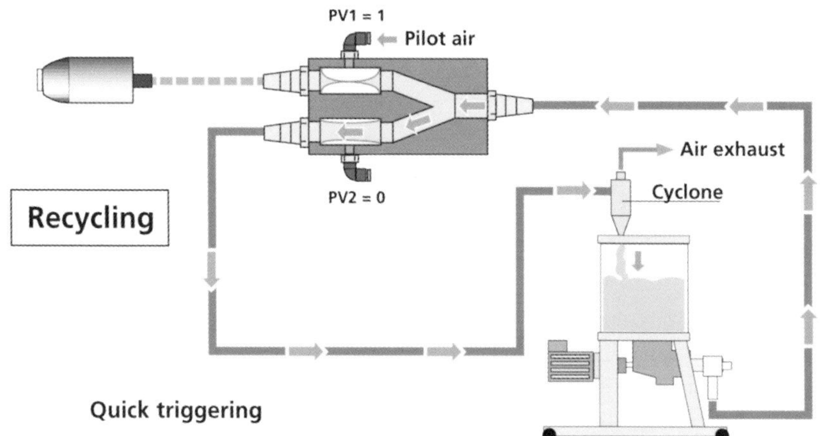

Fast operating powder feed system

The mono-cyclone unit recovers powder by centrifuging it from the air stream and depositing it in a container at the bottom of the cyclone. If the air flow is correct in the ductwork, the system will be self-cleaning. This is obviously very useful indeed when the time taken for colour changes is considered.

Powder can be recycled from the base of the cyclone through a sieve and back to the powder supply feed to the application device.

The ability of a mono-cyclone to recover oversprayed powder can be improved using specially designed multi-cyclone systems. Even in this case many of the fine particles are lost into the final filtration system.

The configuration of either type of cyclone in terms of length and diameter is geared to the capture and separation of specific particle sizes of powder. Multi-cyclone units naturally improve the recovery efficiency. The space requirement is not normally much greater than for a mono-cyclone system, although the cost obviously goes up. The extra cost will soon be recovered by the increased efficiency.

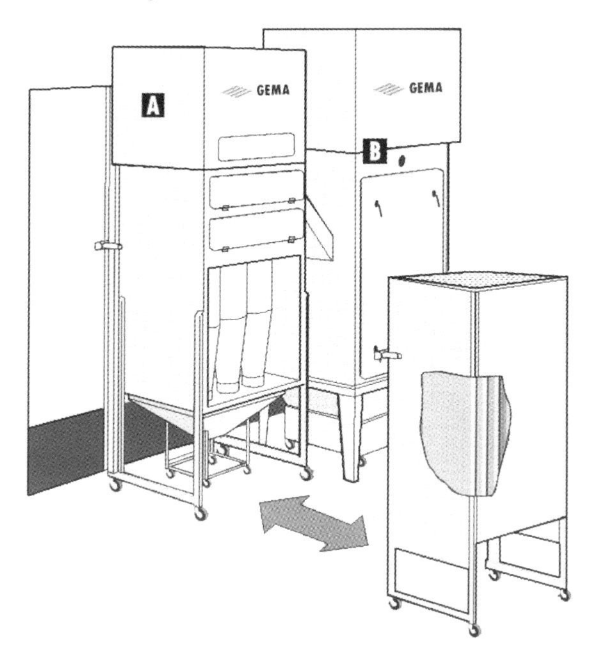

Multi-cyclone system with the option to use a fixed colour cartridge unit

Filters

In dedicated single-colour operations a system using large filter units containing non-woven bags is ideal for the recovery of powder.

Their efficiency is often quoted as more than 99.9% and suppliers will confirm that the air extracted from the spray booth and passed through the filter can be safely ducted back to the factory. Unfortunately these units are not easy to clean, but this is not an issue with single-colour operations.

As mentioned, filter units are often placed behind cyclone systems to extract powder the cyclone unit has not removed. In this instance, the filter unit is not cleaned after each colour change and the powder recovered is disposed of to waste.

Filter units are normally programmed so that they clean themselves at regular intervals to keep the airflow constant, either by vibration or using a short sharp reverse blast from a compressed air supply.

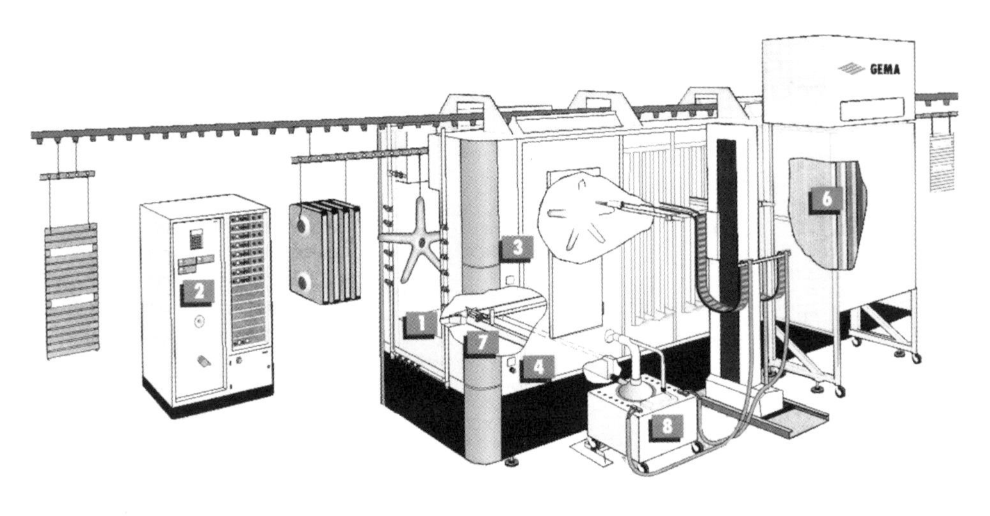

A filter unit

Cartridge filter systems

Many spray booths are now designed with small cartridge-type filters that can recover most particle sizes at greater than 99.99% efficiency. These small units are mounted at the back of the booth and their number will depend on the amount of air and powder involved.

They can be replaced either individually or as a unit when a colour change is required, and give a very flexible approach to the colour change problem.

Booth and cartridge filter system

Horizontal mini-cyclones with cartridge filter assembly adjacent to the spray booth wall

The cartridges themselves are made from a wide range of materials including paper, fabric and non-wovens. It is important to remember that they can have a limited life compared to a cyclone unit.

Many spray booth systems now incorporate a mini-cyclone mounted vertically or horizontally between the cartridge filter units, and this compact assembly enables a colour change within 20 to 30 minutes and prolongs the life of the cartridge assembly.

There is a marked effect on the particle size distribution of the powder in to the recycling process. The cyclone system will tend to recover the coarser particles, and the finer particles will only be retained in the after- filter and are disposed of to waste.

The electrostatic spray process prefers powders of particle size between 15 and 35 microns and with repeated recycling the efficiency of the process will be reduced unless virgin material is added. Eventually defects will begin to occur in the coatings and it then becomes necessary to discard the batch and start again.

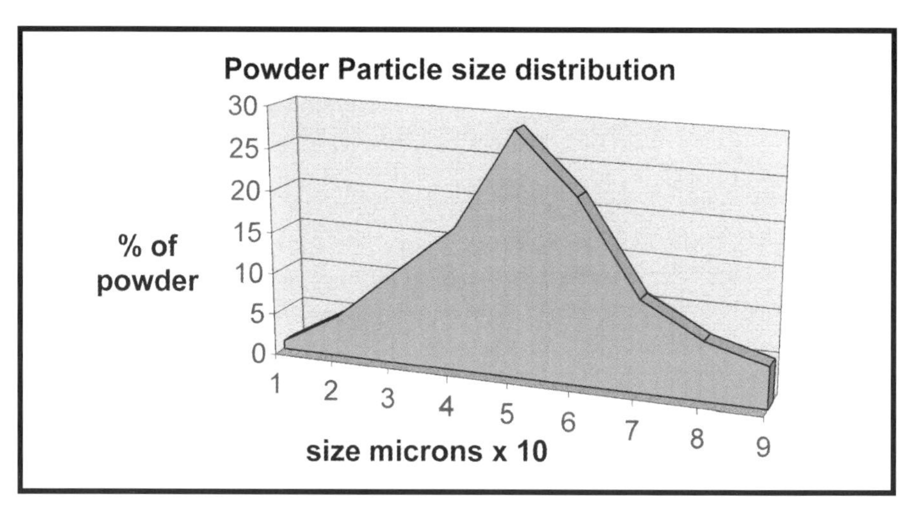

Graph of particle size distribution - virgin coating powder

POWDER RECYCLING

Powder can be recycled from cyclones and filter units for re-use. A venturi device similar to that in the fluidised bed is used for transporting powder to the application device.

Recycling system

The venturi is programmed to move the powder through a powder tube into a mini-cyclone and sieve unit next to the powder feed. The mini-cyclone removes the transport air and the powder falls by gravity into the sieve unit. At this point a proportion of virgin powder is added from the manufacturer's box in a similar manner, the ratio of recovered powder to virgin material being programmed in.

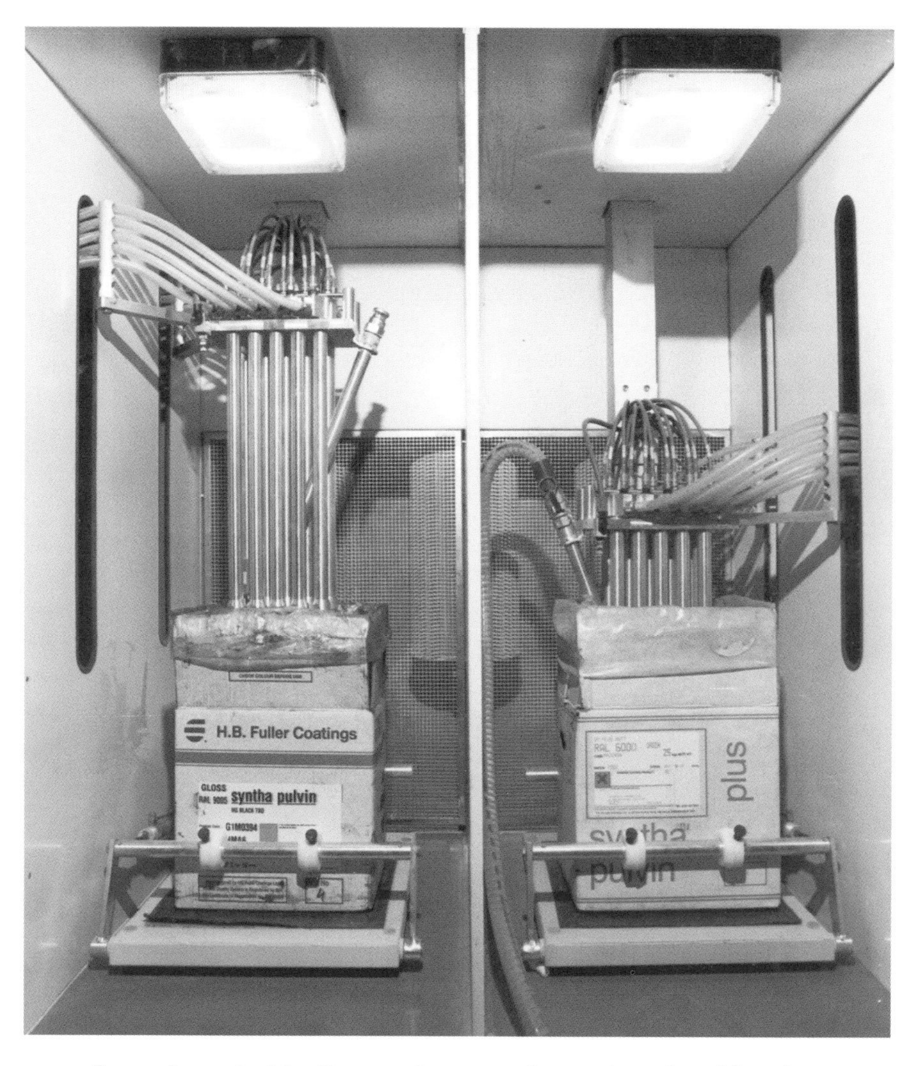

An automated bulk powder supply system to aid colour change time and reduce risk of contamination

CHAPTER VI

HEATING, MELTING AND CURING SYSTEMS

Chapter VI

HEATING, MELTING AND CURING SYSTEMS

The powder coating process depends on heat to allow the coating powder to melt and flow out, and in some cases to cure it. The way this is carried out is an important factor in producing a successful coating.

TEMPERATURE, TIME AND FLOW OUT

The priorities in applying heat to attain the required finish in powder coating are to reach the melt point, enable the correct flow in every case, and then cure it if a thermosetting material. The right temperature and time are critical factors so that each is exactly as the coating formulator intended.

An important point to remember in powder coating is that temperatures are always measured at the substrate, in other words *200°C for 10 minutes* means that the *substrate* has to remain at this temperature for this time to obtain the right result.

The correct heat 'gradient' (rate of temperature build-up) as the component enters the oven after spraying will ensure good flow out, and the component may need to be in the oven for 20 minutes or more to achieve this. For a heavy component it may in fact take as much as 40 minutes to reach 200°C, which means it will have to be at least 50 minutes in the oven.

The term *cure* refers to the conversion of a coating powder polymer into a hard tough film by chemical reaction. In a *thermosetting* coating system the reaction begins once the right temperature is reached.

Thermoplastic materials on the other hand melt and flow out but no reaction or cure takes place. The fluidised bed coating process is therefore particularly suited to this type of coating.

BOX OVENS

These are the simplest form of oven and, as the name implies, consist of a box with doors tailored to suit the particular application. Heating using gas or oil can be either direct or indirect; in electric ovens the air is heated directly.

The heat is transferred around the oven by means of a high-temperature fan located on the cool side of the heat exchanger. Ductwork circulates the air through the heating zone to all corners of the oven. High air velocities are required, at least 4 m/s at the exit of the ductwork. This is insufficient to blow off any of the powder in the case of electrostatic spray application, in which the powder is of course applied to the component cold.

The oven requires ventilating to allow water and by-products of the melting and curing process to escape. In the case of direct-fired ovens the products of combustion must also be removed. The box oven is particularly suited to batch processes.

This type of oven is generally used to pre-heat components for fluidised bed coating. The interior comprises a number of compartments, sequenced either manually or automatically to provide the operator with a component that has reached the required temperature for dipping. The temperature and soak time need to be exactly right to get the required melting and flow in a one-shot dipping operation.

The number of air changes within the oven is indicated by the supplier. It will be far less than for solvent-based paints – another advantage for powder coating.

Legislation and international standards indicate how ovens should be designed to lessen the risk of explosion due to incomplete burning of fuels, and they give advice on the provision of explosion relief panels to provide a degree of safety in an emergency.

It is also important to note that health and safety protection is necessary for operators using ovens as it can obviously be dangerous to enter them while they are still hot. In all cases it must be possible to open the doors from the inside.

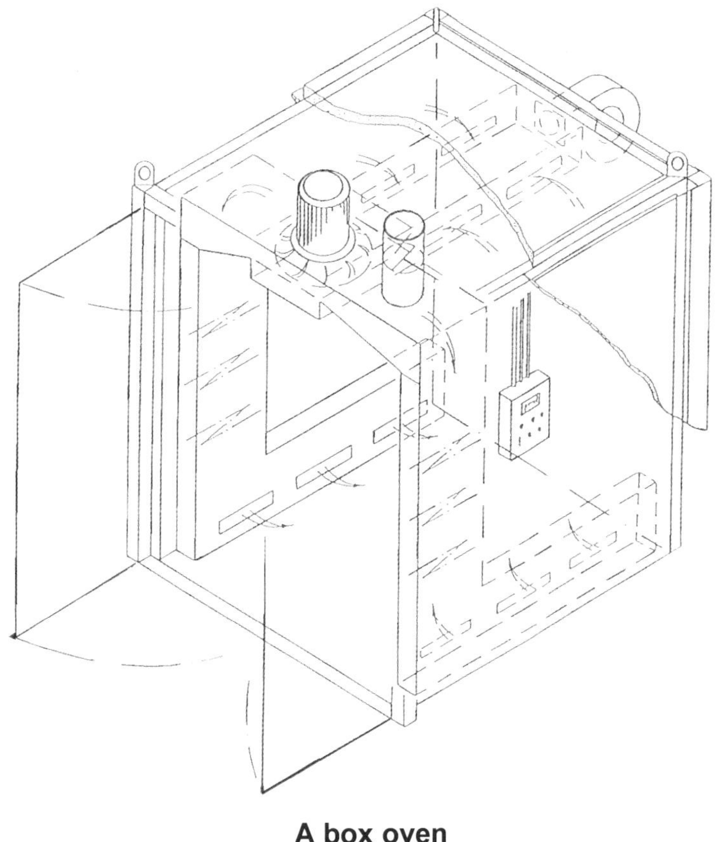

A box oven

CONVEYORISED OVENS

Continuous ovens can be heated by convection and air circulation in the same way as box ovens, or they can use radiant (infrared) heat.

Air circulation ovens can be designed to be in-line unit or multi-pass. In both cases the entrance and exit must be designed to reduce heat loss from the interior. It is normal to add a vestibule at the entrance and exit and incorporate heat seals formed by the movement of air across the silhouette.

The conveyor system passes directly into the oven and it must be constructed from suitably heat resistant components. Oils and greases used for maintenance must also be selected with this in mind.

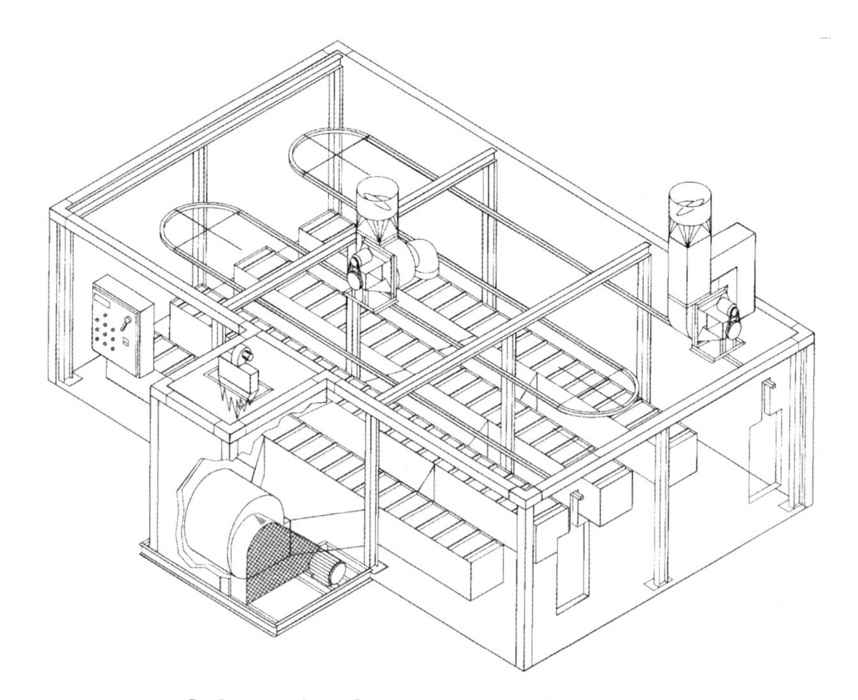

Schematic of conveyorised oven

A multi-pass conveyorised oven

CONVECTION OVENS

The most familiar convection oven is known as a 'camel-back'. It has open ends somewhat lower than the central section which is at the highest temperature. It can be heated directly or indirectly by hot air, generally using gas.

In this type of oven heating and cooling of the articles takes place more gradually than in other types and this has several advantages for powder flow. Its disadvantage is however that the rise and fall of the conveyor tends to limit the size and shape of components that can be handled.

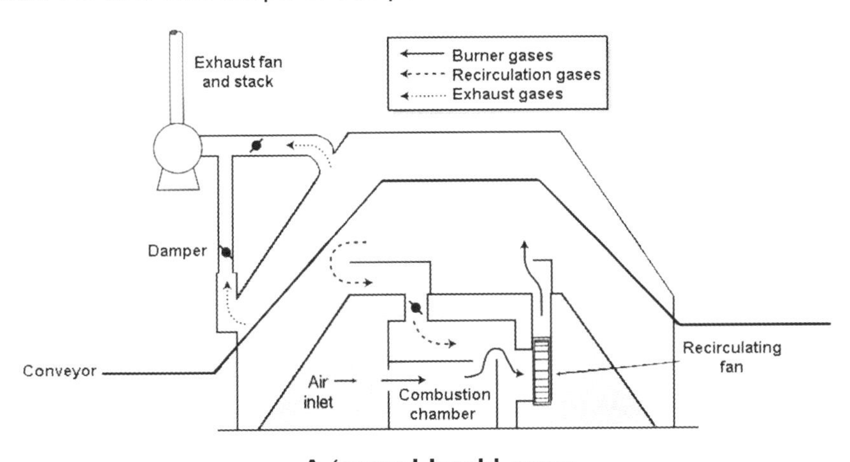

A 'camel-back' oven

In all types of hot air recirculating ovens, the heating, melting and speed of curing speed are affected by the following factors:

- *Heat-up rate.* This is the time taken for the component to reach the required temperature, which may be from cold in the case of a box oven or as it enters the already hot oven in a conveyorised system.
- *Weight of the substrate.* A heavy dense component will absorb a lot of heat before its temperature rises to the right level for flow out and cure to take place.
- *Air circulation within the oven.* It is important to have uniform conditions throughout the body of the oven.

- *Heat losses*. While the body of the oven will be well insulated, a cooling effect can occur if the air circulation around the entrance and exit to a conveyorised oven is disturbed, and performance will suffer as a result.

RADIATION HEATING

Infrared ovens may be heated either by gas or electricity. Gas ovens are provided with a series of panels heated by gas flames impinging on the backs. These act as dull emitters of radiant heat and objects passing between them reach a high temperature very quickly. The time required for heating is very much longer than for convection ovens, but radiant heat is not as good for reaching screened areas and for articles of complicated shape.

Catalytic gas infrared ovens behave similarly.

In an infrared oven some heat is absorbed by the component but the powder coating will absorb the majority.

Electric infrared ovens are composed of a series of heating lamps backed by reflectors, often arranged to form a tunnel. The principle of operation is otherwise the same as for gas ovens.

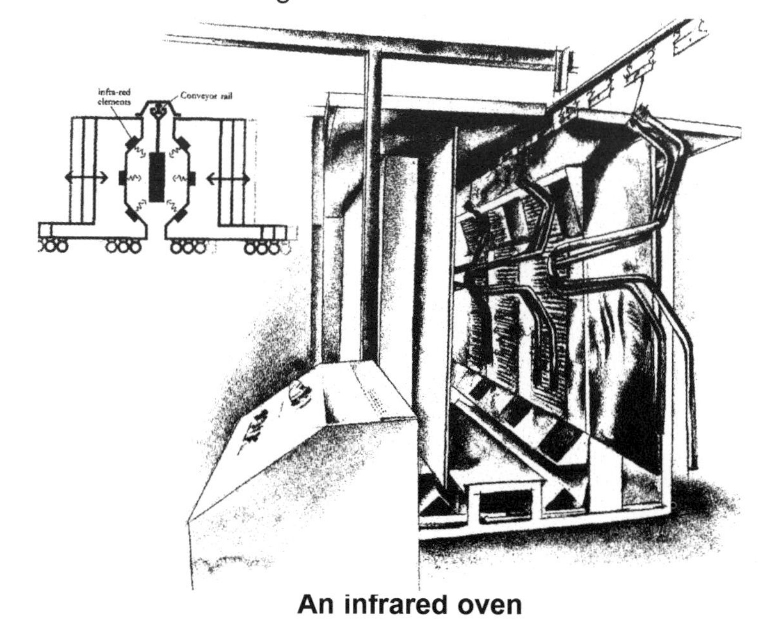

An infrared oven

In this type of oven the air is often recirculated in the direction opposite to the normal flow. This reduces energy consumption and gives more even curing on complicated components.

The melting and curing speed of infrared ovens is affected by several factors:

- *The wavelength of the infrared energy*. Short-wavelength infrared is too fierce for powder coating and is not used. Medium-wave is more suitable – in fact it is so effective that the distance between emitter and component and the time within the heating zone have to be watched carefully. Long-wave infrared is slower and more suited to longer warm-up and cure times.

- *Shape of the components*. Infrared travels in straight lines, and flat panels of sheet metal are cured rapidly because heating is uniform and the powder coating film is of uniform thickness. Complex shapes require an oven able to make use of the hot air generated by the infrared emitters, and are helped if the component is rotated as it heats up. There is a danger of overcure if the component is too close to the infrared source.

- *Conductivity of the component*. Heat conducted through the component is an important factor with awkwardly shaped metal articles. Aluminium conducts better than steel. On the other hand, thick sections of 50 mm or more, or non-metallic components, are cured more quickly in an infrared oven than in the hot-air recirculating type because the coating reaches melting and curing temperature before a significant proportion of the heat has been conducted into the component.

- *Condition of the metal surface*. Bright surfaces reflect infrared heat whereas dull surfaces tend to absorb heat into the metal. This then heats the metal and in turn causes the powder coating to heat up even in areas which infra-red cannot reach directly, clearly good for complicated items.

- *Coating thickness and colour*. On thick coatings care has to be taken to limit the power intensity, or else scorching and degradation may take place. Zoned control can be used to speed up the melting and curing process.

Infrared ovens for pre-heating prior to curing

It can be a big advantage if components are pre-heated using infrared emitters before passing into a hot-air recirculation oven. The infrared section is placed immediately before the opening to the main oven and raises the temperature of the component quickly, reducing the lag time in the oven before the coating begins to flow and then cure. This has the effect of extending the length of the oven without a corresponding increase in floor space. It can also prevent contamination of powders by reducing the space between dissimilar colours on a conveyor system.

OTHER TYPES OF OVEN

There are also some speciality ovens employing induction, ultraviolet and electron beam curing – at the moment these are not very widespread but are very effective in reducing energy consumption.

Induction heating

This is primarily used to melt and cure coating powders applied to metal wire and steel tubes. Each system is designed for the particular application depending on the type of component, the powder (thermoplastic or thermosetting) and the speed of melt and cure required.

Ultraviolet curing

This will be the next step forward in powder coating development. Cold ultraviolet curing enables specialised coating powders to be applied to heat sensitive substrates such as plastics.

A number of lines are in operation around the world. The steps in the process are these:

- Apply coating powder to a conductive plastic substrate.

- Melt the powder using carefully controlled infrared emitters, preferably incorporating a temperature feed-back system, for no more than a few minutes.

- Cure the powder using ultraviolet radiation. Cure is obtained within seconds, but since this is a 'line of sight' process, only fairly flat panels can be cured at present.

The process is quick and does not involve much heat, so the substrate temperature does not rise above 100 or 110ºC.

Electron beam curing

It will be interesting to see what impact this will have on powder coating.

CHAPTER VII

POWDER COATING versus LIQUID PAINT

Chapter VII

POWDER COATING versus LIQUID PAINT

The decision whether to start a new finishing process and move to powder from a liquid paint system is not always easy. Each case is individual and has its own special features.

The key issues are:

- Cost
- Quality
- Health, safety and the environment.

COST

Raw materials

Liquid systems are inherently wasteful, as only a fraction of the raw materials is actually used. The solvent is merely there to aid application and is evaporated off afterwards. As an illustration, it might well happen that out of 100 litres of paint 50 litres is lost to atmosphere as solvent during application and stoving.

Secondly, the application process may only be 60% efficient, so that only some 30 kg, depending on its specific gravity, will actually form the coating on the component. Overspray is lost to the system and the user has the added burden of paying to dispose of whatever sprayed material remains unevaporated.

Third, there will be wastage in mixing, as there is often unused material left over as well as solvent used for cleaning.

The best way to assess these data is actually to measure the raw materials used over a given period, including wastage, against the number of components produced including rejects and re-working. This will give an estimate of the real cost of the liquid paint used for a given number of components. Don't forget to add costs of waste disposal!

By comparison the relative quantity of powder used can be assessed on a comparable basis and the corresponding cost per component worked out.

A second cost calculation including a typical figure for recycled powder might also be made.

Film thickness is not important in itself, as the coatings must be compared at a coating level giving an acceptable performance to the customer.

Processing costs

Pre-treatment is similar whether a liquid system or a powder coating is used, so this is not a factor.

Powder coating has the advantage that there is no heat loss from powder spray booths and air consumption will also be lower. On the other hand, colour changes will be slower and time must be taken to ensure that the application quality is achieved first time as re-work is more difficult with powder.

Fluidised bed application of coating powders is particularly low in cost.

Spray booths used for liquid systems will exhaust large quantities of solvent-laden factory air to atmosphere, so heating bills are higher and the booths need more maintenance. The cost of disposing of solvent-containing waste is high in comparison with powder, although waste powder will also have to be disposed of from time to time.

Stoving, melting and curing costs

Whichever method of application is used powder coating is always expensive in terms of its heat requirement. 200°C is fairly typical, whereas liquid systems are rarely cured above 120ºC and modern two-pack materials need no more than a force-dry at 60ºC. Heating and heat loss are therefore a higher cost with powder.

A simple calculation can be used to work out this cost:

- Temperature increase required.
- Component weight, number and specific heat.
- Heat losses from the oven, including wastage to atmosphere.

Equipment and other costs

For a true comparison it is necessary to include the effect of the capital cost of the equipment used. This can be done by comparing the weekly depreciation on a straight-line basis.

Maintenance requirements may also be quite different for the two techniques.

QUALITY

It is normally assumed that one coat of powder is more than enough for most applications but this is sometimes not the case. One coat of powder is in fact not much better than one coat of paint at the same level of pre-treatment.

Modern architectural performance standards on non-ferrous substrates can be satisfied using powder coating, assuming regular inspection and a limited amount of remedial work. Similar standards are not, however, beyond the scope of liquid paint.

If a paint system is applied over mild steel with a zinc phosphate pre-treatment, consisting of a two-pack epoxy primer followed by a two-pack polyurethane finish, then a corrosion resistance of 1000 hours' salt spray (ASTM B117) may well be achieved at a minimum of 75 microns dry film thickness.

By way of contrast, a polyester powder coating applied at 75 microns over a similar substrate with a zinc phosphate pre-treatment may often not achieve even 500 hours' salt spray. The comparison may therefore be less one-sided than expected.

HEALTH, SAFETY AND THE ENVIRONMENT

There is not much doubt that powder coating is a cleaner, safer and more environmentally friendly process than liquid paint, even if water-based paint is used.

The use of solvents is a burden on the environment and often a health hazard to operatives. There is also the risk of fire, as well as the hazard of storing flammable materials. Many liquid materials are toxic and contact with the skin can cause dermatitis. Liquid systems are also generally messy.

Powder can more easily be controlled but shares some of these hazards, for instance toxicity and the risk to operators of dermatitis and from exposure to dust.

CHAPTER VIII

POWDER COATING IN PRACTICE

Chapter VIII

POWDER COATING IN PRACTICE

The design of the component, the material to applied and the process chosen are key issues in successful coating but experience dictates how easily this will be accomplished.

This chapter gives a number of practical hints and suggestions that may be useful to the first-time applicator and provides a summary of the information given earlier.

METALS

Ferrous and non-ferrous materials were the first to be considered for powder coating, using the fluidised bed process. The technique for these substrates is therefore relatively well developed and understood.

Fluidised bed coating needs careful selection of pre-heat temperatures, bearing in mind the different thickness of metals in the components being coated. Some components with heavy sections will require to 'soak' in the heat, using the retained heat to slowly melt the powder after dipping.

It should be remembered that porous castings, as well as poor quality and incomplete welds, are always liable to produce film defects.

Porosity of castings is not normally a major problem except on cooling, when the coating material can be sucked into the cavities. It is possible to seal casting defects by dipping into a high temperature primer compatible with the powder. An epoxy-phenolic type is useful as this also promotes adhesion.

The electrostatic coating of porous castings can also be a problem as the porosity provides air pockets which 'blow' on heating and curing.

Components can be 'soaked' in the oven at temperatures above the curing point, then allowed to cool somewhat before applying the finish.

When corrosion is a problem, for instance on ferrous substrates, a two-coat powder system can be used. The substrate can be grit blasted or phosphate conversion coated, and then a zinc-rich powder primer applied. A thickness of 40 – 50 microns is advisable.

When this is followed by 50 – 60 microns of a polyester powder it is possible to exceed 1500 hours' salt spray (ASTM B117). The final coat can be applied easily using reduced current and voltage thanks to the conductive primer coat.

GLASS

The powder coating of glass is nearly always carried out by the electrostatic process, though fluidised bed coating has been used in the past on glass components for protection.

To obtain the correct conductivity of the glass so that it can accept a charge from the application device two main methods can be used:

- A conductive coating sprayed on to the glass using water or organic solvent as a carrier will function well but may have disadvantages in environmental terms.

- The component is pre-heated to 60 - 80°C. The heat allows the charge to dissipate at the surface of the glass making it behave as a partial conductor and allowing powder to be deposited. It is worth noting that different types of glass may need a variety of techniques to be tried before a satisfactory result is obtained. Some materials accept a positive rather than a negative charge.

Powder coatings can give both a decorative and a protective result on glass.

WOOD

Wood is a porous material and special powders and techniques are required to enable it to be powder coated successfully. Additionally, the powders must of course be cured at low temperature to avoid degradation of the wood.

The moisture content of the wood has to be carefully controlled, as too little will prevent the powder being applied and too much will ruin the coating.

MDF

Coating of Medium Density Fibreboard is growing fast. It must either be pre-heated or treated by the manufacturer with a conductive layer to enable powder to be applied. This can be included by the fibreboard manufacturer.

A preheat temperature of 60 – 80°C is normally sufficient to allow good coating with a low-cure powder.

There is certainly potential for infrared melting followed by UV curing to give high productivity on flat board.

PLASTICS

The powder coating of plastics is a growing application. Its potential in future may become very important, though there is increasing interest in other developments such as 'in-mould' techniques using powders.

The key issue is the temperature to be used in the powder coating process, namely:

- The temperature the substrate is able to withstand without deforming or degrading.

- The cure temperature of the powder.

The conductivity of the plastic may also be a factor, though this is relatively easy to build in at the compounding stage by the producer.

OTHER MATERIALS

Rubber

Many inorganic powders can be applied to rubber using the electrostatic process, for example talc sprayed on as a release agent to prevent tackiness in handling.

Both organic and inorganic powders can similarly be applied to moulds and other components as release agents.

Textiles

Powder coating can be used to stiffen the interlinings of garments and other similar items. The coating used is generally a thermoplastic material. Low temperatures are used for melting and the application is often carried out by letting the powder flow off a charged blade mounted above the textile moving underneath.

Food products

Electrostatic spraying can be used for applying granular substances and powders such as sugar and flour, and food colourings and flavourings, to a wide variety of food products.

CHAPTER IX

MANAGING QUALITY

Chapter IX

MANAGING QUALITY

Getting the coating right first time must be the aim of everyone involved in the application process.

CONTROL OF QUALITY IN POWDER COATING

Many companies have now adopted methods of measuring and analysing the quality of their processes with the target of achieving zero rejects. Unfortunately it is only the bigger powder coating plants that have so far done this and smaller and medium companies are tending to lose out as a result.

The effect of this is that many powder coaters are wasting money due to poor planning, insufficient process control and not really knowing why rejects are happening. This in turn leads to more reworking, higher costs and in the end unhappy customers.

The adoption of the international standard ISO 9000 by many companies means that they have clearly defined the quality standards to be met in all parts of their organisation and have undertaken to ensure that these standards are met at all times. In the production area this has meant a greater emphasis on training operatives, precise measurement methods and routine inspection.

Ironically, most of the well controlled engineering work going into the production of metal components is finally covered with a layer of liquid paint or a powder coating, as the customer appeal of many products is linked to how they look rather than the quality of their manufacture. The industrial coating scene is now demanding a change of attitude, bringing back a greater emphasis on quality. The new term *surface engineering* expresses very well what is going to be required in future and is of great significance for powder coating. There is a sense in which the next generation of Finishing Shop Managers will need to become 'Quality Engineers'.

THE 'FMECA' TECHNIQUE

A new approach is required to controlling quality in the powder coating industry. No car or domestic appliance manufacturer would ever dream of starting a new project without looking carefully at what potential disasters could be lurking around the corner in the run-up to the product launch. Experience, instinct and common sense are the traditional ways of handling this and there is really no substitute. There are however one or two formal techniques that are useful to back them up, one of which is known as Failure Mode Effect and Criticality Analysis (FMECA). Don't be put off by the name – it can be applied to powder coating very simply and effectively.

The aim is to look carefully at the components to be coated and the processes to be used, and then identify all the possible problems that could arise. The next step is to gauge the effect of each of these on output, quality and cost and then finally to provide ways of reducing the chance of these problems or failures taking place.

In the powder coating process this procedure can have a dramatic effect. It needs of course to be tailored to each unit and then used to encompass all the processes taking place, the potential problems, their causes and effects, and finally the benefits of the actions proposed in relation to their cost.

The steps in this technique are quite simple:

1. Make a note of the components to be powder coated, the type of coat ing to be applied and the method to be used.

2. Decide what possible failures could occur in terms of the end-user's requirements, finish quality, performance and fitness for the intended purpose. You need to list every possibility imaginable.

3. Carefully assess in each case what the cost of each failure would be and what effect it would have on the whole operation. Each should

be given a score of 1 to 10 depending on how seriously you rate them, 1 being not serious and 10 very serious.

4. Using this information it is now possible to list all the problems and the effect of each in order of priority.

5. Use all the information gathered to highlight the corrective actions required, agree who is to be responsible and calculate the pay-back for the expenditure made. The return may be a saving in time, an improvement in output or purely financial. A completion date is also written in.

The technique can be adapted for the powder coating process by adding the refinements shown in the example:

- Listing all the possible causes.
- Scoring the probability of the failure occurring.
- Assessing and scoring the seriousness of each.
- Scoring how visible the failure would be in advance of the completion of the coating process.

The three figures are then multiplied together for each potential failure and the result used to decide how critical it would be. An action plan is given in the final column.

The FMECA calculation takes time to work out the first time but is very worthwhile. It is best to look at each of the problems and their cause several times and to be flexible and broad in your view of them. It is particularly important to discuss the list with all the people involved, including the operators and this will then give you a more complete understanding of the problems, actual or potential.

The one thing that has been found in practice, particularly with reference to powder film defects, is the need for good housekeeping. This not only applies to keeping the work area clean and free from dust and clutter but also to maintenance operations. Housekeeping can be taken to include the routine inspection, test and repair of manual guns, powder feed equipment, booths and recovery systems. Conveyors and hooks require frequent checking for earthing, the effectiveness of filters, powder feed venturis and associated equipment. In practice these are not always properly carried out.

The philosophy of the FMECA approach is so basic that your first reaction will probably be that it is really no more than common-sense. Experience shows however that when it is put into practice, time is dedicated to audit, analyse and take corrective action, and these are procedures that realistically do not happen in the day to day pressure of managing the business. The effect can be virtually guaranteed in cost savings in materials, energy, personnel and management time.

A simple FMECA chart may look like the one on the page following.

Component or process assesssed	Problem or defect	Effect on operation Quality, time, money	Possible causes List all possiblities	Seriousness 1 - 10 1 = seldom 10 = often	Seriousness 1 - 10, 10 = serious	Visiblity before completion of coating 1 - 10, 1 - visible	TOTAL Multiple previous 3 figures	Corrective action and time for completion
Control box	Pinholes	Rework, Rework	Poor welds	5	8	10	400	Set standard for welds - within one week
Frame	Wrong colour	Recoat	Information control	2	10	1	21	Make sprayer responsible for checking paperwork - immediate
Panel	Thin coating	Recoat	Operator technique process control	5	8	5	200	Training provide film thickness gauges within 1month

Quality standards need to be set for all operations. Written procedures are required for the process and work instructions supplied to all those involved. It is not difficult to establish measurements and inspection standards for the product finishing process; most of it can be simplified, leaving out the jargon, and using easy measurements involving simple instrumentation.

The main issue that comes up repeatedly with these new principles is the need for training time. In-house training, however, has to include not only technical subjects but also instilling a sense of ownership and responsibility in the operator involved. What is an acceptable attitude for a machine tool operator has equally to be accepted by the finishing shop operator if quality standards are to be met.

Feedback is then required to measure success, and if a failure occurs it can be used in further refining the FMECA assessment.

Auditing quality

ISO 9000 has become a word-wide standard for quality management. Many companies are proud to have gained accreditation and know that this painful and time-consuming process has been well worth the effort. They recognise its value, both in their dealings with other companies and also in doing things better inside their own company too.

In principle ISO 9000 is simply a matter of stating in the form of a manual how every operation should be carried out and then ensuring that this is what actually happens in practice. It also has a lot to do with auditing, recording, calibrating instruments and the traceability of raw materials. It has surprisingly little to say about intrinsic quality levels as such, in other words its emphasis is mainly on maintaining *consistent* quality. Few people in the company can escape; when an external inspector calls he might well ask a middle manager for his department's training records and will certainly raise an eyebrow if it shows blank since his last visit.

To fulfil the aims of this standard everyone has to be aware that their activities are going to be looked at and monitored. Coupled to this inspection there has to be observation, measurement, and in the case of product finishing an element of visual opinion included.

An important part of ISO 9000 is of course traceability. This means recording part and batch numbers at every stage of production, so that if a problem occurs at a later stage, for example a customer complaint, it will be possible to find out quickly which batches of components were involved and the lot numbers of the powders and other raw materials. Secondly, the job sheet should record any other factors that could be helpful in case of trouble, such as the condition of components to be coated, the actual heating and curing cycle used and the final film properties. Even the temperature and humidity in the finishing shop can later be helpful in investigating a failure.

Statistical process control

When information on product consistency is recorded and analysed it gives the opportunity to apply statistical process control (SPC). The object is to take some of the subjective judgements out of the equation and to replace them with information coming from statistical analysis of the process variables. In other words it should be possible once enough information has been gathered to discover which factors depend on one another and which are most important (and equally which don't really count) in achieving product quality and consistency.

First a simple checklist is produced of all the factors of interest that need to be recorded and then the job process sheet is adapted so that these are systematically recorded day by day. For example:

- Components to be treated:
  - Substrate material (ferrous or non-ferrous metal, wood, plastic, etc.).
  - Type or configuration (casting, fabrication, moulding, extrusion).
  - Thickness (maximum and minimum).
  - Condition (state of welds, cleanness, corrosion, surface defects such as pinholes and porosity).
- Pre-treatment:
  - Process (mechanical or chemical).
- Coating materials:

- Powder (chemical type, supplier, lot number).
- Colour and gloss specified.

- Finishing shop conditions:

  - Temperature, humidity.

- Equipment used:

  - Type (fluidised bed, electrostatic).
  - Equipment used, if more than one.

- Initial assessment of film:

  - Thickness (before curing).
  - Visual opinion.

- Curing or drying:

  - Time, temperature, cure rate, etc.

- Final film properties. Numerical values for:

  - Thickness.
  - Colour and gloss.
  - Adhesion.
  - Hardness and flexibility.

- Rejects and complaints. Number and type by batch:

  - Film thickness.
  - Contamination.
  - Surface defects.

This information can be used at two levels. Firstly it can be looped back into the FMECA assessment in an effort to reduce defects, and secondly it can be used in SPC to study how final film variations are linked to individual process variables, using a simple computer programme available for this purpose.

Adopting a quality assurance culture

For ISO 9000 accredited organisations then a full quality control approach will have been adopted bringing all the finishing shop activities into a properly documented, accountable and traceable system that can be assessed and audited on a regular basis. At the same time all the necessary control and measurement systems will have been introduced.

Other companies prefer the Total Quality Management (TQM) approach. This method uses the dedication and influence of all the senior management percolating down through the organisation to seek a 'zero defect' attitude to *all* operations within the company, from the accounts department through to the shop floor. The cost of failure is recorded and corrective action is instigated to reduce it. The real essence of this method is to bring a sense of responsibility and motivation to the employees without the rigid regime of the ISO 9000 quality system.

A change in outlook is fundamental to the whole operation and this is always a major challenge. The key elements will have to be:

- Willingness from the top to make attitudes change.
- Building motivation and team spirit.
- Training.

The final point, training, is the key in the powder coatings industry. For example, few finishing managers and supervisors understand or have much knowledge of the materials they use. Operators who work in the area know even less. Few trades expect their employees to be able to carry out their duties competently without a knowledge of the materials they are working with.

Operators are generally self-taught and it is regrettable that few receive formal training in the process of dipping or spraying powder coatings.

Film thickness equipment for powder coatings, for example, tends to be jealously guarded by the Quality Control Manager, who does not trust other personnel to assess their own work and cannot see the benefit of this to the organisation.

The attainment of effective quality control in industrial powder application industries will only be brought about by a radical change in philosophy.

Firstly, the managers responsible for finishing activities must come to accept that higher standards of quality are achievable and worthwhile.

Training programmes must be instigated to give in-depth knowledge to all of those involved in the industry.

CHAPTER X

TESTING OF POWDER COATINGS

Chapter X

TESTING OF POWDER COATINGS

INTRODUCTION

If the performance of a powder coating is to be consistent we have to be confident about the quality of the raw materials we are using. This is particularly the case with coating powders as little can be done later to put matters right if they are in any way unsatisfactory.

How do we know if a coating powder as supplied is up to standard and that it will do what we want of it? There is not a great deal that the powder itself can tell us, and most of the information we need can only come when we look at its behaviour and appearance *when applied as a coating.*

There one or two other principles we should bear in mind. Firstly, it is always best to use a test that gives a number as an answer. This then allows us to decide, perhaps in discussion with the customer, what level is regarded as acceptable for that property. In production we call this value the *pass-fail mark.*

Having said this, however, we must not fall into the trap of not actually *looking* at what we are producing. Maybe our customer will be using a slightly different test from ours to decide whether what we deliver is what he wants. One thing we can be completely sure about is that what he actually *sees* (or what the final user sees) is in the end the most important issue. In other words we have to be on our guard against taking refuge behind numbers; common sense always has its part to play.

Secondly, if we are to make sure that quality standards are always met in production, the testing of samples by standard procedures is essential. Corners must not be cut. As we have seen in Chapter IX, our company's reputation for quality and reliability is tremendously valuable to us as it directly affects the amount of business that comes our way and what prices we charge. It can only be built up painfully over a long period of time – but it can be lost overnight if we get things wrong.

There are a number of quite simple tests that we can use. In general it does not matter too much whether a component is coated by dipping or by spraying, the tests used are broadly similar. There are however one or two differences.

EVALUATION OF COATING POWDER

Incoming coating powders can be checked by sprinkling them on to a heated component, or by dipping the pre-heated component into the powder, to assess colour and surface appearance.

In some critical circumstances the melting point of the powder can also be ascertained by placing it on a metal plate, heating it slowly, and noting the temperature at which it begins to flow. This can be useful when working with new materials in order to set the heat-soak parameters for the component to be coated particularly by the fluidised bed method.

TESTS DURING AND AFTER COATING

The performance of fluidised bed coatings largely depends on the quality of the pre-treatment. The majority of coating powders used are of the thermoplastic type, and if pre-heating is good enough to give an even coating of the required thickness then its performance will also be satisfactory.

Tests are carried out to prove the quality of the coating and are the same for fluidised bed and electrostatically applied coatings. All tests are made by the applicator at the end of the process, apart from visual inspection during processing.

Visual examination may be able to indicate:

- Poor pre-treatment.
- Deviations in coating thickness.

Coating thickness assessment of electrostatically applied powders before heating and curing

It is becoming increasingly simple to assess the coating thickness immediately after spraying. If a customer demands 60 microns of coating and insufficient is applied the product will be rejected. When the applicator applies more than 60 microns then powder costs increase and profit is lost.

A simple test is to check the sprayed coating with a clean finger pulled across the surface. The defect can be locally blown-off and simply recoated. A good operator given the opportunity to experiment with test panels can become very competent and will provide an accuracy of ±10 microns.

Some simple devices are available similar to the wet film thickness gauges used with paint and can prove most effective for a modest investment.

Simple powder Coating Thickness Comb

The high-volume powder coater can have other choices, ultrasonic devices are available that can give a fast non-contact powder coating thickness after spraying to within a few microns and though costing several thousand pounds the instruments will recoup their cost in terms of meeting quality standards and reduced powder usage.

Automatic systems can incorporate these devices and hi-tech laser systems to provide a positive feed back to the coating powder supply system and application devices to maintain the coating thickness within the applicator's defined parameters.

Tests after coating are normally as follows:

Colour

Colour measurement equipment makes assessment easy and allows an instrumental tolerance to be used. If it is not available, the assessment may be carried out by eye under lamps conforming to BS 950 Part 1 (e.g. Philips™ 55 tubes) at an intensity of 2000-2500 lux, comparing the colour against a panel previously approved by the customer. A high intensity of lighting makes the assessment much easier.

The purpose of using lamps of standard colour is to avoid disagreements with the customer due to *metamerism*, which means that a panel matched using different pigments will always be a mismatch under a light source of different colour. In practice it is not usually possible to know what pigments have been used in the standard panel, and metamerism is therefore a fact of life. The most troublesome light sources are tungsten lamps but serious problems also arise with natural daylight itself due to its variability. Disagreements are also possible if observers at either the customer or the supplier (or both) are colour blind, a problem affecting almost one in ten of the male population.

Little can be done if the coated components are off-shade except to strip and re-coat. The problem can be caused not only by the powder as delivered but equally by overheating or by over-cure or under-cure, and these should be looked at before blaming the powder supplier.

Gloss

This is measured using an instrument known as a gloss-meter, which measures the proportion of light reflected at various angles from the surface of a panel. The angle usually chosen is either 60° or 20°, and this figure is recorded with the results. A figure approaching 100% indicates high gloss, the ideal case being a mirror: a completely matt result would score zero. It is difficult to make a visual assessment, particularly at gloss levels below 40% or if the surface shows 'orange-peel'.

Dry film thickness

A dry film thickness gauge is used for this purpose. The powder coating must be thoroughly cured before this test is performed and the manufacturer's recommendations must be met.

TEST PANELS

As many of the tests used to evaluate powder coatings are destructive, in other words the coated article is destroyed in the test, arrangements must be made for test panels to be prepared at the same time and under the same conditions as the finished component.

These can be destructively tested at the time and if need be a spare panel can be retained for future reference should problems occur.

The tests involved are as follows:

Adhesion

The BS 3900 cross-hatch test is used for this purpose. Six cuts 1 - 2 mm apart are made with a sharp blade right through the coating to the substrate and a further six cuts made at right angles. Adhesive tape (e.g. Sellotape™) is firmly applied to the surface and then pulled off rapidly ('snatched off'). The proportion of coating removed is assessed.

Cure

A solvent such as xylene or methyl ethyl ketone (MEK) is rubbed on to the cured surface using cotton wool. This is best carried out by one operative to reduce variation in pressure and speed of rubbing. The test is not very precise but is valuable for picking up inadequate cure.

It is very useful if a datalogging instrument is available to record the way the component heats up in the oven to ensure that the heating and curing take place properly. Much can be learned about the oven, the heating characteristics of the component, and its final performance. A history can be built up which will show how the performance of the coating is related to the way it has been cured.

Flexibility

Coated test panels are bent around a mandrel of given diameter to assess coating flexibility. This gives an indication of the degree of cure, as well as any change in the performance of the coating from batch to batch.

TROUBLE-SHOOTING AND OTHER QUALITY ISSUES

Air quality

Powder coating application needs air quality to be controlled and checked regularly. Air used to convey coating powder within the system must be dry and free of particles and oil. Good filtration with a refrigerant dryer in the system will normally meet the required standards.

The dew point of the air supplied to the unit should always be at 4°C or below and there should be less than 0.1 ppm of oil and no particulate matter above 0.3 microns.

Trouble-shooting

Defects can usually be assessed by visual inspection. Often a 50x magnifying glass and a sharp blade will be enough to find out whether defects are occurring within the coating or on its surface, or whether they arise from the surface of the component.

It is particularly important to establish this in the case of pinholes and contamination.

The charts given next will help in investigating and rectifying problems.

FLUIDISED BED COATING TROUBLE-SHOOTING CHART

WHEN DOES THE PROBLEM OCCUR?	WHAT IS THE PROBLEM?	WHAT CAN CAUSE IT?	PROBABLE SOLUTION
AFTER DIPPING	Unmelted coating powder	Insufficient temperature	Increase oven temperature
	Poor flow	Insufficient stored heat	Increase time in oven
	Pinholes	Porosity	Lower oven temperature
			Improve quality of base component
	Uneven coating	Different thickness of metal	Test for optimum time-temperature relationship
ON COOLING	Pinholes	Porosity	Lower temperature of component before dipping
	Differences in surface aspect	Inconsistent cooling	Do not water quench

ELECTROSTATIC POWDER TROUBLE-SHOOTING CHART

WHEN DOES THE PROBLEM OCCUR?	WHAT IS THE PROBLEM?	WHAT CAN CAUSE IT?	POSSIBLE SOLUTIONS
ON APPLICATION	Poor powder attraction:	Bad earthing	Clean hooks and jigs Check earth continuity
	in corners	Equipment problem	Service equipment
	too thin	Faraday cage effect	Lower voltage
		Low voltage	Increase voltage
		Insufficient spray time	Spray component longer
	Pock marks in powder	Poor earthing	Clean hooks and jigs Check earth continuity
		Too thick	Lower voltage
		Back ionisation	Clean hooks and jigs Check earth continuity Lower voltatge
		Voltage too high	Check equipment
	Poor powder flow	Powder damp	Use another box Change storage area
		Poor fluidisation	Increase fluidisation
FINAL COATING	Pinholes and craters	Dirty substrate	Clean the substrate
		Poor pre-treatment	Check pre-treatment process
		Contaminated powder	Check powder
		Back ionisation	Clean hooks and jigs Check earth continuity Lower voltage
	Contamination in film	Contaminated powder	Clean down and change powder
		Dirty coating area	Clean coating area
		Contamination in the oven	Clean the oven Check the powder
	Orange peel	Coating too thick	Apply less powder
		Back ionisation and/or voltage too high	Clean hooks and jigs Check earth continuity Lower voltage

FINAL COATING (cont.)	Too thick	Too much powder applied	Apply less powder Lower voltage Lower Spray time
	Poor adhesion	Dirty substrate Poor pre-treatment Insufficient cure	Examine the cleanliness of the substrate Check the pre-treatment process Check time and temperature of subsstrate in oven Check the powder
	Brittle coating	Overcured	Check time and temperature of substrate in oven
	Soft coating	Insufficient cure	Check time and temperature of substrate in oven Check the powder

Appendix I

HEALTH AND SAFETY AND ENVIRONMENTAL REFERENCES

GENERAL

As well as being controlled by the Health & Safety at Work Act, the application of coatings is also controlled by a series of Health & Safety Executive recommendations. These are constantly being changed with new requirements being introduced.

COSHH

The Control of Substances Hazardous to Health legislation requires the recording of all hazardous materials, safety equipment and training. Specialised publications relating to the application of coatings and the use of flammable materials can be obtained from:

- HMSO Publications Centre
  PO Box 276
  London
  SW8 5DT
  Tel: 0171 873 9090 (orders)
  Tel: 0171 873 0011 (enquiries)
  Fax: 0171 873 8200 (orders)

- HSE Books
  PO Box 1999
  Sudbury
  Suffolk
  CO10 6FS
  Tel: 01787 881165
  Fax: 01787 313995

Substances in powder form always tend to be hazardous to health and should be handled with due respect. Precaution should be taken in both storage and in use, particularly with dust created during spraying. An example of the literature available is:

Control of Exposure to TGIC: HSE Information - Engineering Sheet No 15

SAFETY OF ELECTROSTATIC EQUIPMENT AND POWDER APPLICATION

Modern equipment is inherently safe. Most incidents are caused by poor earthing of the sprayer, the components or other objects within the area of influence of the electrostatic charge.

There are now several standards relating to powder coating and new ones keep appearing.

The most widely used advice in the UK is:

- Code of safe practice: Application of thermosetting coating powders by electrostatic spraying, published by the British Coating Federation.

Address: British Coatings Federation Limited
James House
Bridge Street
LEATHERHEAD
Surrey
KT22 7EP

Telephone: 01372 360 660
Fax: 01372 376 069

International Standards are as follows:

- BS EN 50050: 1986 - Part 1: 1987
  Specification for hand held spray guns and associated apparatus

- BS EN 50053: Part 2: 1989
  Specification for selection, installation and use of hand-held powder spray guns (energy limit of 5 mJ) and associated apparatus

- BS EN 50177: 1997
  Automatic electrostatic spraying installations for flammable coating powder

A draft European Standard in preparation:

- Spray booths for the application of organic powder coating materials - safety requirements

As far as environmental considerations are concerned, many publications stipulate the need for registration and authorisation under certain circumstances to carry out potentially polluting processes involving coatings.

These are published in the form of Secretary of State's Guidance Notes:

PG6/9	Manufacture of coating powder
PG6/23	Metal and plastic coating
PG6/31	Powder coating

It can be seen that PG6/9 and PG6/31 are particularly relevant to powder coating.

STANDARDS RELATING TO APPLICATION AND TESTING

- BS ISO 1461
  Hot dip galvanised coaitngs on fabricated steel

- BS 1722: Part 16
  Specification for organic powder coatings to be used as a plastic finish to wire fences

- BS 3900
  Methods of test for paints

- BS 4479
  Design of articles that are to be coated

- BS 5664
  Solventless polymerisable resinous compounds used for electrical insulation

- BS 6496
  Specification for powder organic coatings for application to aluminium, etc for architectural purposes

- BS 6497
  Specification for powder organic coatings for application to hot-dip galvanised steel for architectural purposes

- BS EN ISO 4618
  Terms and definitions of coating materials

Appendix II

GRAPHS AND FORMULAE

THEORETICAL COVERAGE OF A PAINT

There are three basic criteria for calculating the theoretical coverage of paint:

1. How much solid paint is there per litre of material (powder has 100%).

2. What is the dry film thickness required after drying and / or stoving (evaporation).

3. What is the transfer efficiency if the application process (paint overspray cannot normally be recovered for re-use.

Thus coverage in square metres per litre = 1000 multiplied by the Volume Solids (VS) divided by the dry film thickness (DFT) required and then divided by the process transfer efficiency (TE).

For a 50% volume solids paint, a dry film thickness of 25 microns and an application transfer efficiency of 50%.

$$C = \frac{1000 \times \text{Volume solids}/100 = 20}{\text{Dry film thickness}}$$

20 x TE

20 x 50/100 = 10 sqm / litre

THEORETICAL COVERAGE OF A POWDER COATING

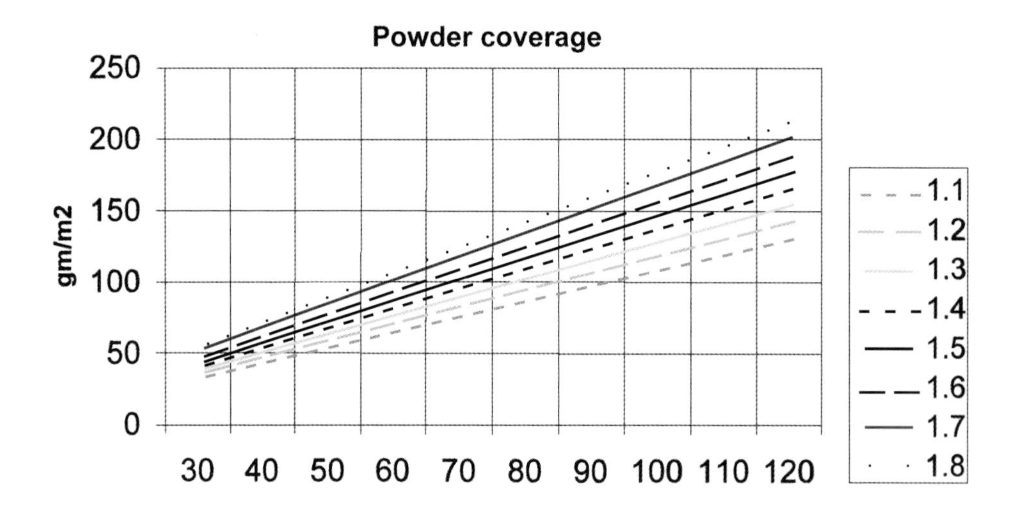

SCHEMATIC AUTOMATIC SPRAY GUN MOVEMENT ON CONVEYORISED SYSTEMS

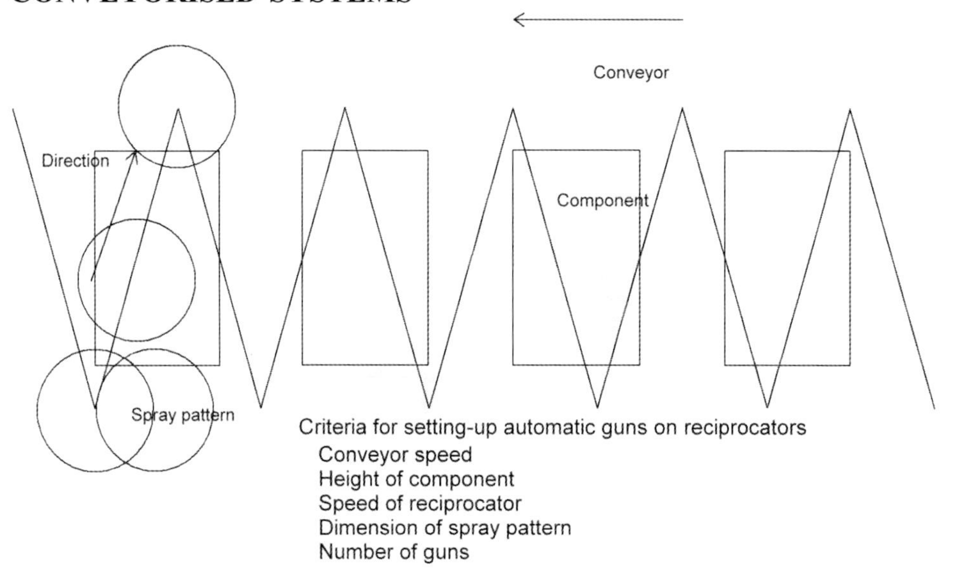

Criteria for setting-up automatic guns on reciprocators

- Conveyor speed
- Height of component
- Speed of reciprocator
- Dimension of spray pattern
- Number of guns

The aim is to obtain even coverage over the entire area at minimum reciprocator speed < 30m/s

PARAMETERS RELATING TO POWDER FLOW

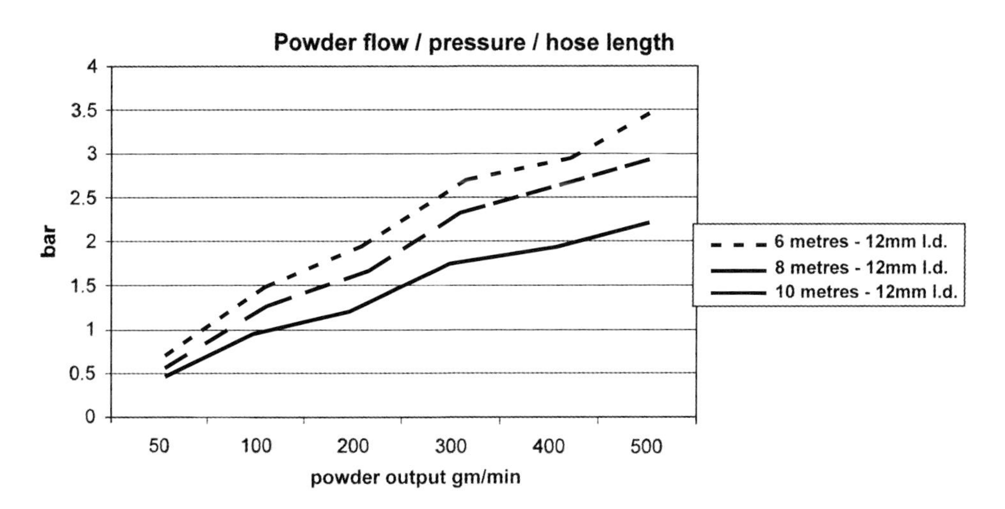

AIR FLOW, VELOCITY AND VOLUME

Air flow is measured by multiplying velocity by cross sectinal area.

Thus velocity (m/sec) multiplied by cross sectional area (m^2) will give the volume of air moved in m^3 / sec.

LOWER EXPLOSION LIMIT OF COATING POWDER

The maximum concentration of powder present in a powder booth is calculated as:

$$C = M / V$$

C = the concentration of the powder in the booth
M = is the amount of powder emittted from the guns per unit of time
V = the volume of air extracted from the booth per unit of time

It is normal to use the maximum output of the guns and the minimum possible amount o fair extracted from the booth.

The lower explosion limit is measured in g/m^3. It is generally agreed unless specifically defined by the coating powder manufacturer that the IEL (lower explosion limit) should not be greater than 10 g.m3.

Standard coating powders have a typical LEL range of 20 g.m^3 - 70 g.m^3

GRAPH SHOWING HOSE CHARACTERISTICS RELATED TO LENGTH AND DIAMETER AND THEIR EFFECT ON POWDER FLOW

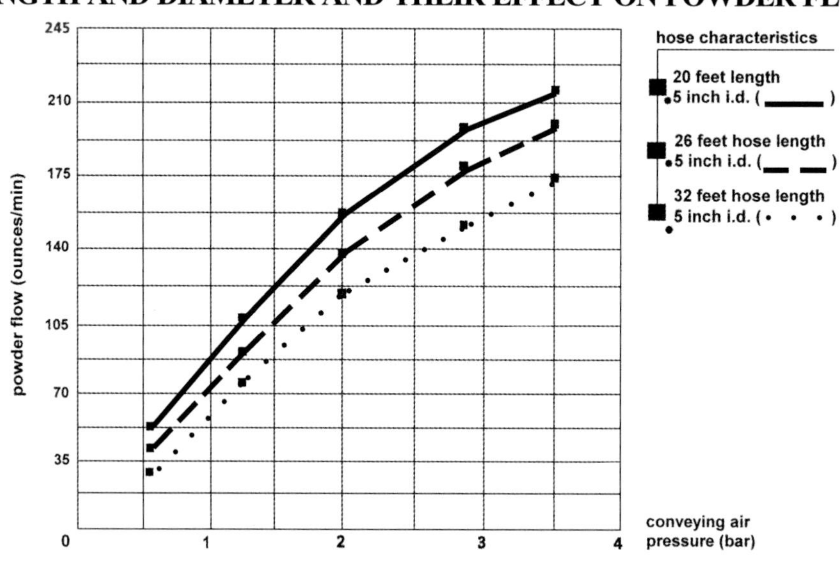

OVEN COSTS

Calculating the running costs of Ovens:

1. Obtain the cost of energy:
   Gas
   Electricity

2. Monitor energy consumption either by fitting local meters or monitoring energy usage when other equipment has been isolated.

3. A general energy usage will probably be based upon:

   - Start-up
   - Steady use

4. Heat losses will occur from:

   - The structure - walls, floor, roof
   - Exhaust ventilation : 2 to 4 times more on solvent based paint ovens
   - Heat loss through conveyor openings - conveyorised ovens
   - Component heating and associated carriers and jigs

5. Thus when considering ovens and cost difference from paints to powder they will be:

   - Higher temperature with powder
   - Lower exhaust emissions with powder

Appendix III

GLOSSARY

BACK IONISATION. A condition occurring during electrostatic application of powder in which an excessive build-up of charged powder particles limits further powder being deposited on the substrate. It can reverse the electrical charge of the surface layer of powder producing pockmarks.

COMPATIBILITY. The ability of powder coatings from different sources or of different compositions to be mixed and applied without visible or mechanical differences in the cured film or application properties.

CORONA CHARGING. The process of inducing a static electric charge on powder particles by passing the powder through an electrostatic field generated by a high-voltage device. The charge is normally applied at a needle or sharp edge.

CURE PARAMETERS. The time-temperature relationship required to properly crosslink and thus cure a thermosetting powder coating.

DRY BLEND MANUFACTURE. A process used in the manufacture powder coatings in which the components are blended without subsequent extrusion and melting.

EDGE COVERAGE. The ability of a powder coating to flow over, build and adhere to sharp corners, angles and edges.

ELECTROSTATIC DEPOSITION. A technique of charging powder so that it is deposited on to an earthed substrate.

ELECTROSTATIC FLUIDISED BED. A method of charging powder in a fluidised state so that it can be deposited on to an earthed component.

ELECTROSTATIC SPRAY. A method of spraying electrostatically charged powder so that it is deposited on an earthed component.

FARADAY CAGE EFFECT. A condition that inhibits the electrostatic application of powder particles to a specific area due to its geometric configuration such as an angle, where powder would have to travel into an enclosed area despite the presence of nearer conductive surfaces.

FILM FORMATION. The forming of a continuous film by melting powder particles and coalescing them by the application of heat.

FLOCKING, FLOCK SPRAYING. A method of applying a second layer of powder or other material by spraying on to the sticky surface of a component that has previously been powder coated or given an adhesive layer. If it has had a previous powder coating applied it is heated to slightly above its melting point.

FLUIDISED BED. A box-shaped container in which powder is suspended in a continuous stream of air. Pre-heated objects may be coated by dipping them directly into powder fluidised in this way. A fluidised bed may also be used to transport powder materials to other types of application equipment such as spray guns.

FUSION. The melting and flow out of individual coating powder particles under the influence of heat to form a continuous film.

GEL TIME. The interval required at a given temperature for a powder to be transformed from a dry solid to a gel-like state.

EARTHING. Connection of an object to a neutral source (earth) so that it cannot support an electrostatic charge.

FERROUS. Based on iron or one of its derivatives such as steel.

FLOWABILITY. A term used to describe the ability of the dry powder to flow uniformly or to be continuously poured from a container at a steady rate. This is important as the better the flow the less the air required to transfer the material to the application device.

IMPACT FUSION. The tendency of finely divided powders to melt and aggregate in the application equipment due to the formation of local hot-spots.

INTERCOAT ADHESION. The ability of a coating to adhere satisfactorily to a previously applied film.

ION STRIPPING. The process of removing ions from the air stream at the nozzle of a corona-charged spray gun using earthed needles. It is designed to avoid back ionisation.

LOWER EXPLOSIVE LIMIT (LEL). The lowest concentration of powder particles suspended in air that can be ignited by a spark or flame.

MELT POINT, MELTING POINT. The temperature at which a finely divided powder will begin to melt and flow.

MICRON. A unit of length amounting to one millionth of a metre (10^{-6} m).

NON-FERROUS. Based on a metal not containing iron or steel.

PASSIVATION. A treatment designed to make a surface inactive and less prone to corrosion.

PARTICLE SIZE. The average dimension of a coating powder particle. Although irregular in shape, in most calculations they are assumed for simplicity to be spherical.

PLASMA DEPOSITION TECHNIQUE. A method of applying powder using compressed gas in which the powder is melted in a flame before it impinges on the surface of the component.

POCKMARKS. Small pitted defects in the surface of a coating.

POWDER CLOUD CHAMBER. An enclosed booth containing a cloud of charged powder through which a component can be passed and coated.

POWDER COATINGS. Protective or decorative coatings formed by the application of a coating powder to a substrate and fused into a continuous film by the application of heat or radiant energy. Coating powders are finely divided particles of an organic polymer containing pigments, fillers and other additives. They remain stable on storage under suitable conditions.

QUENCHING. Cooling a hot component by dropping or dipping it into water.

RECLAIMED POWDER. Powder remaining unused ('oversprayed') during the application process that is collected and recovered for re-use.

RECOVERY. The process of removing oversprayed powder from the air before reclaiming it for re-use.

STORAGE STABILITY. The ability of a powder coating to remain unchanged in its physical and chemical condition during storage.

STRIPPING. Removal of a previously applied coating from a component or an item of handling equipment.

SURFACE APPEARANCE. The overall smoothness and gloss of a powder coating film and the presence of surface defects.

SURFACE PROFILE. The microscopic appearance of a surface in elevation, measured in the relative height of the peaks and troughs.

THERMOPLASTICS. A group of resins used in powder coating that on heating will soften and melt without curing. This enables them to be re-melted on subsequent heating. Examples are PVC, nylon, and polyolefins.

THERMOSETTING RESINS. This group is the counterpart of thermoplastic resins. They normally melt on heating but on further heating they undergo crosslinking to form a cured film that cannot be re-melted. Also known colloquially as 'thermosets'. Examples are expoxies, polyesters and acrylics.

TRANSFER EFFICIENCY. The percentage of sprayed powder that is actually deposited on to the object being sprayed. The balance is known as 'overspray'.

TRANSPORTABILITY. The ability of a powder coating to be carried by an air stream through tubing and ducts designed to transport it to the application device. It is related to 'flowability'.

TRIBOCHARGING. The process of creating a static electric charge on powder particles by friction against a non-conductive material.

VENTURI. A tube containing a specially designed constriction which produces a pressure differential.

VIRGIN POWDER. Coating powder as supplied by the manufacturer as opposed to powder that has been reclaimed or recycled.

VOLATILE CONTENT. The percentage of volatile matter lost by a powder on heating under prescribed conditions.

WRAP-AROUND. A characteristic of powder coatings applied by electrostatic spraying to coat parts of the component not in a direct line with the spraying device.

Appendix IV

FURTHER READING

The Science of Powder Coating: Chemistry, Formulation and Application Vols. I and II; SITA Technology.

Phosphating and Metal Pre-treatment, D B Freeman; Woodhead-Faulkner.

The Phosphating of Metals, Werner Rausch; ASM International.

Electrostatics and its applications, A D Moore; John Wiley and Sons.

User's Guide to Powder Coating, Darryl L Ulrich; Society of Manufacturing Engineers

The Powder Coating Handbook, The Powder Coating Institute.

ETSU Guides

GG15 Vapour degreasing

GG50 Cost-effective paint and powder coating – materials management

GG51 Cost-effective paint and powder coating – surface preparation

GG52 Cost-effective paint and powder coating – coating materials

GG53 Cost-effective paint and powder coating – application technology

GPG 260 Optimisation of industrial painting and powder coating

GPG 271 Selecting and specifying new paint and powder curing ovens

These are available in the UK, free of charge,

from: ETSU
Harwell
Didcot
Oxfordshire
OX11 0RA

Telephone: 01235 436 747
Fax: 01235 433 066

INDEX

A

Abrasive pads 40-41
Acrylic based powders 134
Adhesion 9, 19, 21-24, 36, 37, 187
Air flow 136, 147, 149
Alcohols 50
Application of Powder coating 69-115
    advantages/disadvantages/alternatives 74-75
    colours 70
    costs 73
    electrostatic spray 71
    Electrostatic conveyorised powder coating 82-83 * **See Electrostatic conveyorised fluidised bed powder coating**
    Electrostatic spraying 82-113 * **See Electrostatic spraying**
    flock spraying 82
    fluidised bed 73, 75
    Fluidised bed coating 75-81 * **See Fluidised bed coating**
    production factors 72
    specifications 71
    substrates 67, 70-71
    thickness 72
Assessment of component 13-17
    bolts and screws 16
    crimping 15
    design 13-17
    edges 13-14, 16
    hinges 17
    laser 14
    substrates 11
    threads 16
    welds 14
Automatic applicators 113-115
Automatic booths 128-130

Automatic electrostatic spraying equipment 107-111
Automatic water wash 132, 134

B

Bolts and screw holes 16
Box ovens 148-149
Brushing 40

C

Chemical conversion coatings 36, 38, 53-56, 59, 62-63
Chemical etches 24
Cleaning 28-30, 38-64, 123-124, 131, 134
Coating suitability 8, 17
Colour 6
Colour change 6-7, 130-134
Colour testing 186
Component position 9
Contamination 9, 18, 22-23, 25, 36-38, 51-52, 131-134, 135, 136
Convection ovens 151-152
Conveyor loading 27-28
Conveyorised fluidised bed coating 82
Conveyorised ovens 149-150
Conveyorised tunnel systems 47
Corrosion 6, 9, 37, 52-53, 65, 162
Corrosion resistance 161
Crimping 15
Curing systems 148-155
Cure tests 187-188
Cyclone recovery 137-139

D

Degreasing 19, 35, 42
Design 13-17, 122-123, 124-128
    manual booths 124-128
    spray booths 122-123
Dipping 26
Dust 25, 36

E

Earthing 27
Edges 13-14, 16
Effluent treatment 63-64
Electrostatic conveyorised fluidised bed powder coating 105-106
  bed coating 105-106
  flock spraying 106-107
Electrostatic powder spray guns 99-105
  Corona 100
  Tribo 101-105
  - cleaning 104
  - nozzles and extensions 102-1043
  - test kit 104
Electrostatic powder coating 4
  atomiser 5
Electrostatic spraying 83-115
  air assisted spray gins (Corona charging) 108-110
  automatic air assisted spray gun (Corona charging) 108-109
  automatic applicators 114-115
  automatic electrostatic spraying equipment 107-115
  benefits and disadvantages 84-85
  box units 95-97
  corona charging 86-89
  epoxy or polyester powders 88
  Faraday cage 89
  flock spraying 106-107
  fluidised bed coating 105-106
  fluidised bed hopper 93-95
  injectors and venturis 97-99
  ion stripping 86-88
  manual electrostatic application 91-92
  nozzles and extensions 102-103
  operating principles 91-105
  powder bells 111
  powder discs 111-113
  powder feed systems 93
  PTFE 89
  spray guns 99-105
  spraying equipment (automatic) 107-111

theory of electrostatic powder coating 85-86
Tribo charging 89-91
Tribo disk 108-109
Evaluation of coating powder 184
Extrusion 19

F
Failures 21-23
adhesion 9, 19, 21-24
delamination 22
interfacial chalking 22
scab corrosion 22
weathering 22
wetting contact theory 23
Ferrous metals 19, 20, 165
Film thickness testing 184-185
Filters 134, 135, 139-142
Flame spraying 4
Flexibility 9, 188
Flock spraying 106-107
Fluidised bed cleaning 28, 29
Fluidised bed coating 75-82
conveyorised fluidised bed coating 82
dipping 78
equipment 75-81
masking 79
porous membrane 76
post heat and curing 81
problems 78-81
- preheat temperatures for Rilsan™ Nylon[11] 79
thermoplastic and thermosetting powders 76
water quenching 81
wire coating 80-81
FMECA Technique 172-180

G
Galvanised components 19-20
Glass 18, 20, 37, 44, 166
Glossary of Terms used in Powder Coating 203-207

Gloss tests 186
Graphs and Formulae 197-201
Grit blasting 20, 35, 41-44, 166

H
Handling components during and after pretreatment 24-31, 59
chemical stripping 28
conveyor loading 27-28
conveyorised systems 59
dipping 26
earthing 26
fluidised bed cleaning 29
high boiling solvents 30
high temperature adhesive tapes 30
holding components 25-26
hooks and fixtures for dipping 26, 27, 28
incineration 28-29
jigging devices 25, 26, 27
masking and shielding 27, 30-31
mechanical stripping 30
Health and Safety 30, 119-120, 122, 126, 126-128, 129-130, 162, 193-196
Health and Safety and Environmental References 193-196
COSHH 193
electrostatic equipment and powder application 194-195
Standards relating to application and testing 195-196
Heating, melting and curing systems 147-155
box ovens 148-149
convection ovens 150-152
conveyorised ovens 149-150
other types of oven 154-155
- induction heating 154
- ultraviolet (UV) curing 154-155
radiation heating 152-154
- features 152
- factors affecting melting and curing speed 153
- infrared ovens 154
temperature/time and flow out parameters 147
Heat up rate 151

Humidity test 35

I

Incineration and oven 28-29
Induction heating 154
Infrared ovens 154
Interfacial chalking 22
International standard - IS 9000 167-168

J

Jigs 25, 26, 27

L

Laser cutting 14, 19
Lower Explosive Limit 119-120
Low volatility esters 50

M

MDF 18, 167
Manual electrostatic application operating principles 91-93
Manual spray booths 125-128
Masking and shielding 9, 27, 30-31
Mechanical pretreatment methods 35, 40
Mechanical stripping 30
Metal substrates 165-166
Minimum film weight 6
Moisture content 21

N

Non ferrous metals 19, 161, 165
Nut shell crushed 42
Nylon polymers 79

O

Operating principles for manual electrostatic application 91-93
Ovens 148-152, 201

P

Paint coverage 197-201

air flow, velocity and volume 199
criteria 197-199
- charts and graphs 198-199
- dry film thickness 193
- process transfer efficiency (TE) 197
- volume solids (VS) 197
Lower Explosive Limit (LEL) 199
oven costs 201
Particle size 135, 140, 142
Phosphating 56-59
Planning the powder coating process 7-10
component position and masking 9
design 9
manufacturers specifications 9
type of coating 9
Plastics 7, 18, 20, 24, 37, 44, 167
Powder coating in practice 165-168
component design 165
glass 166
MDF (Medium Density Fibreboard) 167
metals 165-166
plastics 167
other materials - rubber/textiles/food products 168
wood 167
Powder coating process 7-10, 17-24, 36-38
adhesion 9, 19, 21-23, 35, 38
contamination 9, 18, 22, 23, 25, 36, 36-38, 58-59
corrosion 6, 9, 19, 22, 36-38, 40
flexibility 9
passivation 19, 58-59
planning the process 7-10
possible processes 8
suitability for coating 8, 9
thermoplastic powder 7
thermosetting powder 7
Powder coating testing 183-191
evaluation of coating powder 184
tests during and after coating 184-187
colour 186

coating thickness assessment 185
dry film thickness 187
gloss 186
test panels 187-188
adhesion 187
cure 187-188
flexibility 188
trouble shooting 188-191
air quality 188
electrostatic powder trouble shooting chart 190-191
fluidised bed coating trouble shooting chart 189
trouble shooting 188
Powder coating versus liquid paint 159-162
costs 159-161
- equipment 161
- processing costs 160
- raw materials 159-160
- stoving, melting, curing 160-161
Health, Safety and the environment 162
quality 161-162
Powder flow 123, 124
Powder spray booths 119-144
automatic booths 128-130
- requirements 128-130
booth configurations 121
booth selection (required features) 124-130
cleaning 123-124
colour change 130-134
- computerised systems 131
- features/problem solving 131-133
- plant design examples 133-134
design 122-123
Health and Safety 119-120
manual booths 125-128
- adequate air flow 126-127
- electrical safety 127
- features/problems 125-128
- lighting 127
- noise level 127

- operator safety 128
materials of construction 120-122
polymer type change problems 134
powder recovery 136-144
cartridge filter systems 140-142
cyclone recovery 137-139
- particle size 142
- mono-cyclone 138-139
- multi-cyclone 139
filters 139-140, 140-142
powder recycling 142-144
transfer efficiency 135
powder application/recycling and recovery 135
Pretreatment cleaning process for powder coating 9, 18-21, 38-64, 64-65
abrasive pads 40-41
acidic aqueous solutions 51
alkaline aqueous solutions 52
British Standard BS 7079:1989 43
brushing 40
chemical cleaning 44-45
chromate conversion coatings 62-63
conversion coatings 36, 38, 53-61
chromating 53, 54, 62-63
combined cleaning and conversion 59-61
passivation of phosphate layers 18, 58-59
phosphating 53, 54, 55, 56-59, 60-61
conveyorised tunnel systems 47
degreasing (and phosphating) 35, 42, 46-47, 59-61
effluent treatment 63-64
low volatility esters and alcohols 50-52
mechanical treatments 40-52
mould release lubricants 44
power wash cleaning 48-49
shot or grit blasting 20, 35, 41-44
solvents 29-30, 39, 46, 48
spraying (for cleansing) 48
steam cleaning 49-50
Swedish Standard SIS 055900 42-43
ultrasonics 45, 47-48

vapour degreasing 45-46, 51
waterbased systems 50-52
wipe clean 38-40
Pretreatment for powder coating process 9, 18, 19-20, 25, 35-65
adhesion test 35
chemical conversion coatings 36, 38, 53-56
chemical treatments 25, 36. 44-45
degreasing 19. 35, 36, 42, 45-48
glass 18, 20, 24, 37
grit blasting 20, 35, 41-44
mechanical treatments 35, 38-44
non ferrous oxides 37
plastic 7, 18, 19, 24, 37
rust 37
salt spray and humidity 35
static electricity 37
water wash and pressure 35
wiping 35, 38-40
Pretreatments - Other 64-65
electrophoretic primers 65
phosphoric acid washing 64
pigmented etch primers 64-65
primer coating powders 65
Process costs 159-161
Product quality 161-162
Productivity 4

Q

Quality Control management 171-180
FMECA (Failure Mode Effect and Critical Analysis) technique 172-180
adopting a quality assurance culture 179-180
- ISO 9000 & TQM (Total Quality Assessment) 175
auditing quality 176
calculation/chart/assessment 169-172
statistical process control 177-178
steps in technique 172-176
ISO 9000 171-172

R

Radiation heating 152-153
Raw material costs 159-161
Recovery and recycling (of powder) 135, 136-142, 142-144
Rust 37

S

Salt spray test 35
Safety Standards 193, 194-196
Scab corrosion 22
Shot blasting 20, 35, 41-44
Spray cleaning 48
Spray booths 119-144 * **See Powder spray booths**
Static electricity 37
Steam cleaning 49-50
Substrates 7, 13, 18-21, 24, 36, 36-65, 70-71, 161, 161-162
- extrusions 19
- ferrous metals 18, 19, 20, 165
- food products 168
- galvanised components 19-20
- glass 18, 20, 24, 37, 166
- grit blasted 166
- MDF (fibreboard) 18, 37, 167
- non ferrous metals 16, 19, 161, 164
- plastics 7, 18, 20, 24, 37, 167
- porous castings 165-166
- phosphate conversion coated 166
- pretreatment 18, 19-21, 25, 35-64
- rubber 168
- textiles 168
- wood 18, 21, 37, 167
- zinc 18, 19, 20

Surface profile 44

T

Temperature/time 147
Testing of powder coatings 183-191
Theoretical calculations and tables 197-201
- air flow, velocity, volume 199

coverage of paint 197-199
LEL - Lower Explosion Limit 200
oven costs 201
powder flow 199
Thermoplastic resins 7, 144
Thermosetting resins 7, 23
Threads (coating of) 16
Transfer efficiency 135, 136
Trouble shooting 188-191

U
Ultrasonic cleaning 47-48
Ultraviolet curing 7, 154-155

V
Vapour degreasing 45-46, 51
Venturis 97-99, 142-144

W
Waterbased cleaning systems 48-49, 49-50, 51-52
Waterquenching 26
Water wash for pretreatment 35
Weathering 22
Welds (coating over) 14-15
Wetting contact theory 23
Wiping 35, 38-40
Wood 18, 21, 37, 167

Z
Zinc coated substrates 18, 19, 20
Zinc phosphate pretreatment 161